DSP Filters

Electronics Cookbook Series

Acknowledgments

The authors thank Judy Lane and Dan Hoory for reviewing selected chapters and providing invaluable comments and suggestions. They also thank Kim Heusel for his hard work and long hours preparing the final manuscript.

DSP Filters

by
John Lane
Jayant Datta
Brent Karley
Jay Norwood

Electronics Cookbook Series

PROMPT® PUBLICATIONS

©2001 by Sams Technical Publishing

PROMPT© Publications is an imprint of Sams Technical Publishing, 5436 W. 78th St., Indianapolis, IN 46268.

All rights reserved. No part of this book shall be reproduced, stored in a retrieval system, or transmitted by any means, electronic, mechanical, photocopying, recording, or otherwise, without written permission from the publisher. No patent liability is assumed with respect to the use of the information contained herein. While every precaution has been taken in the preparation of this book, the author, the publisher or seller assumes no responsibility for errors or omissions. Neither is any liability assumed for damages resulting from the use of information contained herein.

International Standard Book Number: 0-7906-1204-6
Library of Congress Control Number: 2001088401

Acquisitions Editor: Alice J. Tripp
Editor: Kim Heusel
Assistant Editor: Cricket A. Franklin
Typesetting: Kim Heusel
Indexing: Kim Heusel
Proofreader: Cricket A. Franklin
Cover Design: Christy Pierce
Graphics Conversion: Christy Pierce
Illustrations: Courtesy the authors

Trademark Acknowledgments:
All product illustrations, product names and logos are trademarks of their respective manufacturers. All terms in this book that are known or suspected to be trademarks or services have been appropriately capitalized. PROMPT® Publications and Sams Technical Publishing cannot attest to the accuracy of this information. Use of an illustration, term or logo in this book should not be regarded as affecting the validity of any trademark or service mark.

PRINTED IN THE UNITED STATES OF AMERICA

9 8 7 6 5 4 3 2 1

Table of Contents

Chapter 1 - Introduction 1

Section I - Filter Design Formulas 5

Chapter 2 - Filter Design Basics 7
Low-Pass Filter 7
High-Pass Filter 10
Bandpass Filter 10
Band-stop Filter 12
Peaking Filter 12
Shelving Filter 14

Chapter 3 - Digital Basics 17
Analog versus Digital 17
Sampling Theorem, Aliasing, and Quantization 19
Mathematical Transforms 24
Digital Filter 27
Unit Circle 29
Digital Signal Processing Operations 30
Types of Digital Filters 32
Comparison of IIR and FIR Filters 34
The Biquad Section 35
Returning Back to the Analog Domain 36

Chapter 4 - First-Order Low-Pass Filter 39
Analog Filter Network 39
Digital Filter Network 42
Difference Equation 46

Chapter 5 - First-Order High-Pass Filter 49
Analog Filter Network 49
Digital Filter Network 52
Difference Equation 56

Chapter 6 - Second-Order Low-Pass Filter 59
Analog Filter Network 59
Digital Filter Network 63
Difference Equation 67

Chapter 7 - Second-Order High-Pass Filter 69
Analog Filter Network 69
Digital Filter Network 73
Difference Equation 76

Chapter 8 - Second-Order Bandpass Filter 79
Analog Filter Network 79
Digital Filter Network 84
Difference Equation 90

Chapter 9 - Second-Order Band-Stop Filter 93
Analog Filter Network 93
Digital Filter Network 98
Difference Equation 105

Chapter 10 - Peaking Filter 107
Digital Filter Network 107
Difference Equation 113

Chapter 11 - Shelving Filter 115
Low-Pass IIR **115**
Digital Filter Network 115
Difference Equation 118
High-Pass IIR **120**
Digital Filter Network 120
Difference Equation 123

Table of Contents

Chapter 12 - Cascaded Low-Pass Filter 125
Analog Filter Network 125
Digital Filter Network 131
Difference Equation 136

Chapter 13 - Cascaded High-Pass Filter 139
Analog Filter Network 139
Digital Filter Network 145
Difference Equation 150

Chapter 14 - Cascaded Bandpass Filter 153
Digital Filter Network 153
Difference Equation 161

Chapter 15 - Cascaded Band-Stop Filter 163
Digital Filter Network 163
Difference Equation 171

Section II - Filter Projects 173

Chapter 16 - Introduction 175
Overview 175
Project Outline 176

Chapter 17 - Tone Control 179
Design Requirements 181
Filter Overview 181
Functional Block Diagram 185
Flow Diagram Descriptions 186
Software Description 191

Chpater 18 - 60 Hz Hum Eliminator 205
Design Requirements 206
Band-Stop Filter Overview 207
Functional Block Diagram 208
Implementation 211

Chapter 19 - 31-Band Graphic EQ-I 223
Design Requirements 225
Peaking Filter Overview 226
Functional Block Diagram 228
Flow Diagram Descriptions 230
Software Description 235

Chapter 20 - 31-Band Graphic EQ-II 245
Design Requirements 245
Bandpass Filter Overview 246
Functional Block Diagram 248
Flow Diagram Descriptions 249
Software Description 256

Chapter 21 - 4-Band Parametric EQ 267
Design Requirements 270
Filter Overview 271
Functional Block Diagram 276
Flow Diagram Descriptions 278
Software Description 284

Chapter 22 - Digital Crossover 297
Which Filters Should be Considered in Crossover Designs? 299
Design Requirements 308
Filter Overview 308
Functional Blocks 315
Control Flow Descriptions 317
Software Description 319

Appendix - Odd-Order Filters 335
Digital Low-Pass Filter 336
Digital High-Pass Filter 338
Difference Equations 339

Index 341

References 344

1
INTRODUCTION

Digital filters and real-time processing of digital signals have traditionally been beyond the reach of most hobbyists, due partially to hardware cost as well as complexity of design. In recent years, low-cost digital signal processor (DSP) development boards have been introduced to the market in a price range within the budget of most home electronics enthusiasts. However, complexity of design has still been a major hurdle to the hobbyist, with the result that most DSP development boards have remained in the hands of the design engineer, as well as students enrolled in university engineering programs.

DSP Filters is an attempt to break down this design complexity barrier by means of simplified tutorials and step-by-step instructions along with a collection of audio projects. This book is written in the spirit of the *Active Filter Cookbook* (Don Lancaster, 1982) and *Design of Active Filters with Experiments* (Howard Berlin, 1977). These previous books shattered the complexity barrier of active filter design. In a similar manner, we have chosen to describe digital filter design by presenting design formulas needed to build the digital equivalent of standard audio filters: low-pass, high-pass, bandpass, and band-stop, also including the more specialized peaking and shelving filters. Section I: Filter Design Formulas describes the design and analysis formulas of both analog and the digital equivalents of 14 specific filter types.

The filters described in the active filter design books may be conveniently classified as first and second order. Cascading first- and second-order sections create higher order filters in the appropriate combinations. The particular class of digital filters described in this book includes those that most closely resemble analog filters — *Infinite Impulse Response* (IIR) filters (another name for this type of digital filter is recursive filter). Each class of analog and digital filters is characterized by poles and zeros in the complex *s-plane* (for analog) or complex *z-plane* (for digital).

The goal of Section I is to present the digital filter design formulas and compare them to the equivalent analog formulas whenever possible. No derivations are given with the hope of keeping the technical level constrained. The only requirement for the reader is the ability to understand algebraic formulas involving trigonometric functions.

Section II: Digital Audio Filter Projects describes implementation examples of the various filter types using the C++ programming language, which will further support the mathematical descriptions. Note that Section I can be used strictly as a reference section. Therefore, one strategy for the reader is to skip directly to the Section II projects. The specific design formulas from Section I needed in each of the projects are repeated in Section II for convenience.

Five audio projects are described in Section II. These projects have been chosen for their potential interest to the electronics hobbyist, with special consideration of the usefulness in illustrating the design and implementation of concepts presented in the chapters of Section I.

Introduction

Unlike active filter design methodologies, DSP filter design is primarily a software task. For that reason, the breadboard with collections of op amps, resistors, and capacitors, can be substituted for a DSP evaluation module (EVM) development board. Wire-wrap tools, pliers, and a soldering iron are replaced with DSP tools — the compiler (and/or assembler), simulator, and evaluation module/debugger.

Since most modern DSP development systems begin with a C++ compiler, the authors of this book have chosen to present the projects as C++ programming examples. This is the most general approach and does not lock the examples into a single DSP software instruction set.

In summary, *DSP Filters* has a two-fold purpose:
1. Provide digital filter design formulas, comparing these to the analog filter equivalent whenever possible.
2. Demonstrate these concepts by building, or more specifically, coding audio projects using the C++ programming language.

Section 1

Filter Design Formulas

2

Filter Design Basics

Low-Pass Filter

A *low-pass filter* may, in general, be defined as a filter that permits a signal to be passed with little or no modification, from $f = 0$ Hz up to a *cutoff frequency* f_c, above which the signal is rejected.

Figure 2-1 shows a generalized low-pass *filter specification*. This figure represents the desired output signal level relative to the input

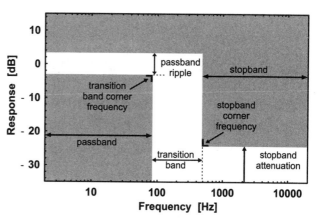

Figure 2-1. Example response specification for a low-pass filter

signal level, over a frequency range (plotted on a logarithmic scale). As shown in Figure 2-1, the range of frequency covered by the unaltered signal up to the *transition band* is known as the *passband*. The range of frequency immediately above the point at which the signal is attenuated below a predefined level is the *stopband*. The level that defines the point at which the stopband begins is based on the requirements of the filter specification, which in turn is dependent on the particular application.

The general design requirements of a low-pass filter can be specified by the following set of parameters:
- *Passband to Transition Band Corner Frequency*, f_1.
- *Passband Ripple*: maximum gain variation (in dB) throughout the passband.
- *Transition Band to Stop-band Corner Frequency*, f_2.
- *Stop-band Attenuation*: minimum filter attenuation throughout the stop band.

An example of a physically realizable filter response, the second-order low-pass filter, is shown in Figure 2-2. The gain, $G(f)$, defined as output level divided by the input level (usually a function of frequency), can be conveniently expressed in *decibels* (dB), as:

$$dB \equiv 20\log(gain) = 20\log\frac{\overline{V}_o}{\overline{V}_i} \quad (2\text{-}1)$$

where $\overline{V}_i$ is the time average of the input, and $\overline{V}_o$ is the time average of the output signal. When the output level is greater than the input level, there is a gain associated with the filter, and the corresponding response in dB is positive, i.e., $G(f) > 1$. If the output is less than the input level, there is an attenuation associated with the filter processing so that the corresponding response in dB is negative, in which case $G(f) < 1$.

The cutoff frequency f_c, is normally defined as the frequency where the gain is reduced to $1/\sqrt{2} = 0.707$ times the passband level (-3dB down from the passband level). As a side note: *most of the filters we will be considering in this text have a flat passband response.*

Filter Design Basics

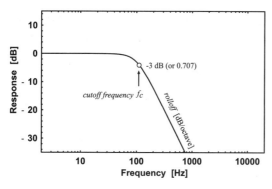

Figure 2-2. Gain response plot of a second-order low-pass filter with cutoff frequency $f_c=100Hz$

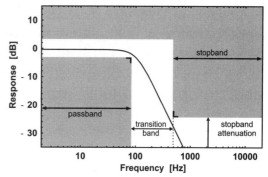

Figure 2-3. Comparing the second-order filter response from Figure 2-1 to the low-pass filter specification of Figure 2-2

The transition band response of most N^{th}-order low-pass filters, when viewed on a dB versus log frequency plot, often decrease linearly with frequency. The rate of decrease is dependent upon the order of the filter and is measured in dB/octave or dB/decade. The rate of decrease is often referred to as the filter *rolloff*. A –6 dB/octave rolloff occurs for each order of the filter. For example, the rolloff of a first-order filter is –6 dB/octave, whereas the rolloff of a second-order filter is –12 dB/octave.

An *octave* is a doubling or halving of the frequency. For example, the octaves above 2 kHz are 4 kHz, 8 kHz, 16 kHz, etc., and the octaves below 2 kHz are 1 kHz, 500 Hz, 250 Hz, etc. A *decade* is a tenfold increase or decrease in frequency. For example, the decades above 1 kHz are 10 kHz, 100 kHz, etc., while the decades below 1 kHz are 100 Hz, 10 Hz, and 1 Hz. A –6 dB/octave is equivalent to –20 dB/decade.

Figure 2-3 shows that a solution to the example filter specification from Figure 2-1 is the second-order low-pass filter of Figure 2-2. The primary goal of most filter design activities is to determine and verify a filter design configuration whose characteristics (cutoff frequency and rolloff) match the filter specifications. Even though this may be the general strategy for most filter design problems, we will be taking a different

DSP Filters

approach in this text. Our target application is audio. It has been accepted that most audio applications require specific filter design characteristics for specific applications (more on this later in the project chapters). Therefore, our filter design strategy in this text is reduced to choosing and customizing a filter whose behavior can be described by a small number of characteristic parameters: filter *type*, *cutoff frequency*, filter *bandwidth* or *Q*, and filter *order*.

High-Pass Filter

The generalized *high-pass filter* is shown in Figure 2-4. The high-pass filter's operation is exactly the opposite of the low-pass filter. This filter rejects a signal below the cutoff frequency and allows the signal to be passed with little or no attenuation above the cutoff frequency. The filter concepts described for the low-pass filter, including rolloff, transition band, and cutoff frequency also apply here for the high-pass filter. Mathematically one can think of this as reversing the direction of frequency.

Bandpass Filter

The *bandpass filter* processes a signal as shown in Figure 2-5. This filter type permits a *band* of frequencies to pass with little or no attenuation and rejects the signal outside the *passband*. The middle of the band is called the *center frequency* f_0. The *band-edge* or *cutoff* defines both the lower f_1 and upper f_2 portions of the band. The cutoff is typically defined to be −3 dB below the passband maximum level. The *bandwidth* Δf is defined as the difference between the upper and lower frequencies: $\Delta f \equiv f_2 - f_1$.

The quality factor Q is inversely proportional to the frequency bandwidth Δf of the gain response curve, and proportional to the center frequency:

Filter Design Basics

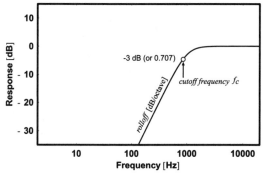
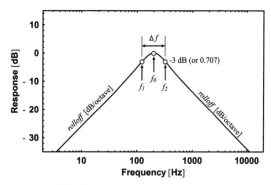

Figure 2-4. Gain response plot of a second-order high-pass filter with cutoff frequency $f_c = 1000$ Hz

Figure 2-5. Gain response plot of a second-order bandpass filter with center frequency $f_0 = 200$ Hz and $Q = 1$

$$Q \equiv \frac{f_0}{\Delta f} \quad (2\text{-}2a)$$

where again, $\Delta f = f_2 - f_1$.

The center frequency f_0 is equal to the geometric mean of f_1 and f_2:

$$f_0 = \sqrt{f_1 f_2} \quad (2\text{-}2b)$$

The *band-edge* frequencies f_1 and f_2 are defined by the following relationships:

$$f_1 = \frac{f_0}{2Q}\left(\sqrt{1+4Q^2} - 1\right) \quad (2\text{-}3a)$$

$$f_2 = \frac{f_0}{2Q}\left(\sqrt{1+4Q^2} + 1\right) \quad (2\text{-}3b)$$

Using Equations (2-3a) and (2-3b), the edge (or cutoff) frequencies in Figure 2-5, with $f_0 = 200$ Hz and $Q = 1$, are: $f_1 = 123.607$ Hz and $f_2 = 323.607$ Hz.

DSP Filters

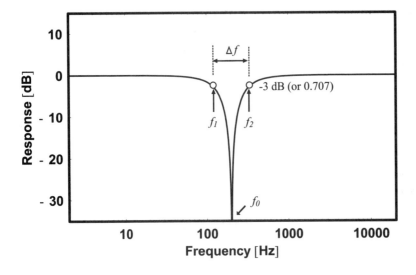

Figure 2-6. Gain response plot of a second-order band-stop filter with center frequency $f_o = 200$ Hz and $Q = 1$

Band-stop Filter

The *band-stop filter* is shown in Figure 2-6. This filter is also known as the *band-reject* filter or the *notch filter*. The band-stop filter's operation is exactly opposite of the bandpass filter: it rejects a *band* of frequencies while allowing frequencies outside of the band to pass with little or no attenuation. As with the bandpass filter, the middle of the band is characterized by the *center frequency* f_0. The *cutoff* defines both the lower f_1 and upper f_2 portions of the band. The cutoff is typically defined to be −3 dB below the level of the passband. The mathematical relationships for the bandstop filter between center frequency f_0, edge frequencies f_1 and f_2, and bandwidth (or Q), are also precisely described by Equations (2-2a) through (2-3b).

Peaking Filter

The *peaking filter* has a characteristic bell shape and is therefore also known as a *bell filter*. Example responses are shown in Figures 2-7a

Filter Design Basics

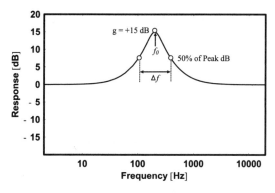
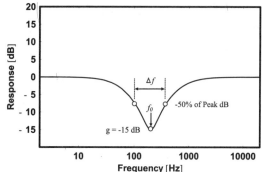

Figure 2-7a. Response plot of second-order peaking filter with $f_0 = 200$ Hz, $Q = 1$, and $g = +15$ dB

Figure 2-7b. Response plot of second-order peaking filter $f_0 = 200$ Hz, $Q = 1$, and $g = -15$ dB

and 2-7b. The peaking filter's shape is qualitatively similar to the bandpass filter with the exception that the *out-of-band* frequencies are left unaltered (0 dB gain) rather than attenuated, as in the bandpass filter case. The peaking filter can either boost or attenuate a band of frequencies while having no effect on the signal frequencies outside the band. As with the bandpass filter, the center of the band is characterized by its *center frequency* f_0.

The most important characteristic of a peaking filter is its boost/cut symmetry. For example, Figure 2-7a shows a peaking filter response where the boost gain at the band center is $g = +15$ dB. The equivalent cut response with a gain value of $g = -15$ dB is shown in Figure 2-7b. Note that these response shapes are mirror images of one an another.

The definition of bandwidth is somewhat subjective in the case of peaking filters. All of the filters discussed up to this point consistently use a −3 dB gain value (0.707) to mark the cutoff or edge frequencies. Since there is no guarantee that a peaking filter is boosted or cut beyond +/− 3 dB, a better convention for bandwidth is the width where the peak is at 50% of the peak value. For example, the peaking filter in Figure 2-7a has a peak gain of +15 dB. Therefore, the bandwidth can be measured at the +7.5 dB level.

Peaking filters usually have a bandpass filter at their core (which is discussed in more detail in Chapter 10). The core bandpass filter has a definite Q value associated with it, as described in a previous section. The relationship between this Q value and the bandwidth of the peaking filter is somewhat complicated. Therefore, we will adopt the convention that the peaking filters we will discuss are described by intrinsic Q value, which is qualitatively related to bandwidth. But we will not attempt to define, quantitatively, the peaking filter bandwidth, using this Q value (refer to Chapter 10 for more on the peaking filter).

Shelving Filter

The low-pass *shelving filter* response, shown in Figures 2-8a and 2-8b, is similar to the low-pass filter discussed previously with the exception that the frequencies above cutoff remain unaltered. As in the case

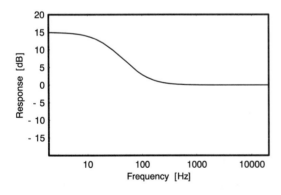 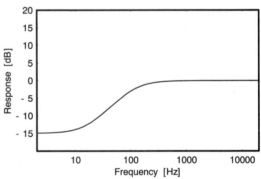

Figure 2-8a. Response plot of a first-order low-pass shelving filter with f_c = 30 Hz and g = +15 dB

Figure 2-8b. Response plot of a first-order low-pass shelving filter with f_c = 30 Hz and g = –15 dB

Filter Design Basics

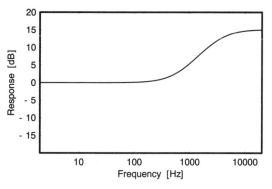
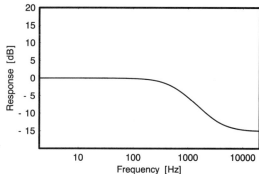

Figure 2-8c. Response plot of a first-order high-pass shelving filter with f_c = 2000 Hz and g = +15 dB

Figure 2-8d. Response plot of a first-order high-pass shelving filter with f_c = 2000 Hz and g = -15dB

of the low-pass filter, a characteristic frequency, or cutoff frequency f_c, determines the point where the response makes its transition from boost or cut to the unaltered filter state (with a response of 0 dB) above cutoff.

Like the peaking filter of the previous section, the shelving filter can either boost or attenuate low frequencies. As with the peaking filter, the shape of the boost and cut modes are mirror images of each other (see Figures 2-8a and 2-8b). In addition, the characteristic cutoff frequency qualitatively controls the point of boost/cut transition. The cutoff frequency f_c used to implement the shelving filter is based on the core first-order filter making up the shelving network. f_c determines the –3 dB point of the core first-order filter.

The shelving filter also works as a high-pass filter, again with symmetric boost/cut modes (see Figures 2-8c and 2-8d). Shelving filters are discussed in greater detail in Chapter 11.

3
DIGITAL BASICS

Analog versus Digital

In Chapter 2 the concepts of some of the most commonly used filter types were introduced. These filter concepts have traditionally been realized in the analog domain by using operational amplifiers typically in conjunction with resistor and capacitor networks. Tone controls, graphic equalization, parametric equalization, rumble filters (for vinyl records), and loudspeaker crossover networks are a few examples of filtering applications that are commonly seen in audio equipment. Filtering in the analog domain has been the mainstay of audio filtering in the past and some of these implementations continue to exist in products even to this day.

There is another, more powerful approach that is quickly gaining in popularity — implementing the same functionality by using digital means via digital filters. The rest of the book focuses on these techniques and this chapter provides a quick tutorial to digital systems, which is considered a requisite precursor to understanding, designing, and implementing digital filters. If the reader is already familiar with the basics of digital systems, he or she may either move to the following chapters or skim through the contents of this chapter as a refresher.

At the outset, it would be interesting to compare and contrast these two technologies and try to gain an understanding of why one is used over another. The main reasons for the longevity of analog implementations is that they are relatively inexpensive to produce, it is an area that is very well understood, and there is a natural inertia against change. What digital technology provides, almost by definition, is a great amount of flexibility. This will become evident in subsequent chapters. However, to illustrate a point, one might consider implementing a reverberation section and a parametric equalizer section for the purpose of processing an audio signal. In the analog domain, these two functions require very different hardware circuits — there is almost nothing in common between these two applications. In the digital domain, however, one may treat a digital signal-processing chip as a black box that is capable of providing both functions with changes to the code that is running within it. That is, the hardware architecture for implementing both functions is exactly the same. This results in the final system design being potentially simpler while offering greater flexibility.

There have been some barriers that needed to be overcome to use digital technology—the initial cost of these systems was high, though they have decreased in price with the passage of time. It has required the services of a software programmer to realize these implementations, which added to the cost of the system. Though this is outside the scope of this book, it should be noted that by merely mimicking analog filters, a digital system is being underutilized. A digital system is capable of creating radically different filter structures that may be more applicable for certain applications.

Sampling Theorem, Aliasing, and Quantization

Audio signals that are encountered in our everyday lives are analog in nature, which is the way we as humans hear sound. Hence, these signals need to be digitized if we are to harness the tremendous processing power and techniques available in the digital domain. A process called sampling achieves this. Essentially, this process is equivalent to someone measuring the amplitude of the audio waveform with a digital voltmeter at a periodic rate and recording this information.

Consider an analog sinusoidal wave in the time domain having a frequency of *f* as shown in *Figure 3-1*. The analog signal is said to be continuous in time and continuous in amplitude, that is, it is a smooth looking waveform. Now, the same signal is shown sampled (or digitized) at differing periodic rates (fast rate and a slow rate, respectively) in *Figures 3-2* and *3-3*. These figures will help to introduce two digital concepts known as the *sampling theorem* and *aliasing*.

It is clear that the continuous time nature of the analog signal has been replaced with a signal that exists only at discrete points in time. This leads one to be intuitively concerned about the fact that information may have been lost in the sampling process. As the dotted line in *Figure 3-2* shows, the original sine wave may be recovered from the discrete time samples, but that is not the case in *Figure 3-3*. The reason is that the Nyquist criterion or the sampling theorem rule was violated. This requires the sampling frequency (f_s) to be greater than twice the highest frequency present in the signal. In the case of *Figure 3-3*, the sampling frequency is too slow. This results in the original signal being undersampled. As the dashed line indicates, these discrete samples are actually interpreted as being a sine wave of a much lower frequency. This phenomenon is known as aliasing—where the original waveform (of a

DSP Filters

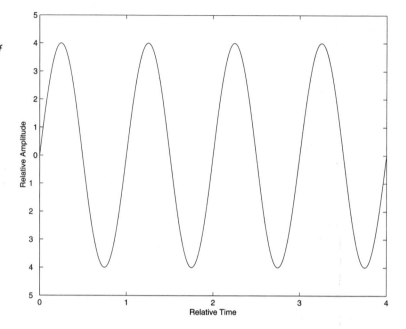

Figure 3-1. A few periods of an analog sine wave of frequency f

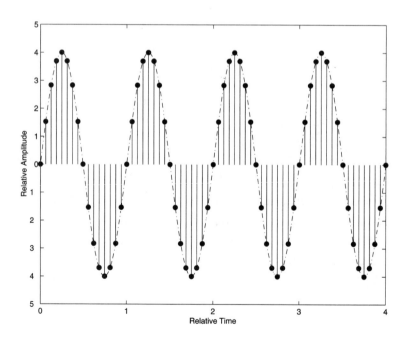

Figure 3-2. Analog sine wave of Figure 3-1 is sampled at a rate that is greater than twice the sine wave frequency. A dotted line joins the samples, thus tracing the original waveform. This shows that no information was lost in the sampling process.

Digital Basics

higher frequency) is interpreted as a signal of lower frequency (that is less than half the sampling frequency).

Hence, violating the sampling theorem leads to the possibility of aliasing. Normal human auditory bandwidth is considered to span the range from 20 Hz to 20 kHz. In order to avoid aliasing, the sampling frequencies used for audio applications need to be greater than twice the audio bandwidth. Some commonly used sampling frequencies are 44.1 kHz and 48 kHz.

It is entirely possible that the incoming audio signal may contain high-frequency harmonics (beyond the audible range) that may cause aliasing down in the audible band. This is strictly a digital artifact that has no equivalent in the analog world. Fortunately, there is a way to prevent this. If the audio signal is passed through an anti-aliasing filter at the front end, then aliasing may be prevented in the digital system downstream from it. This takes the form of a low-pass filter that has been designed to reject signals above a certain frequency.

In addition to being continuous in time, an analog waveform is said to be continuous in amplitude as well. Please note that due to the finite resolution of the hypothetical voltmeter being used to sample the waveform, there is an error introduced when the amplitude value of the sample is recorded. This is illustrated in Figure 3-4, where for simplicity it is assumed that the resolution of the voltmeter is 1 volt. Thus, the maximum error in recording the amplitude is ± 0.5 V. In general, the maximum error is one-half the least significant bit (LSB) of the digital representation. This is because the continuous amplitude signal of the analog waveform would require an infinite number of bits. In reality, only a finite number of bits are used to store a digital signal. The addition of each bit results in the error signal being reduced by half—in other words, the addition of each bit results in the signal-to-noise ratio (SNR) being increased by about 6 dB. By controlling the number of bits of resolution, one may control the amount of quantization error introduced in the sampling process.

DSP Filters

Figure 3-3. The analog sine wave of Figure 3-1 is sampled at a rate that is less than twice the sine wave frequency. The dot-dash line traces the original analog signal. Owing to undersampling of the original waveform, the digital system interprets the samples as a sine wave of much lower frequency (denoted by the dashed line). In this case it is clear that information was lost. This phenomenon is known as aliasing.

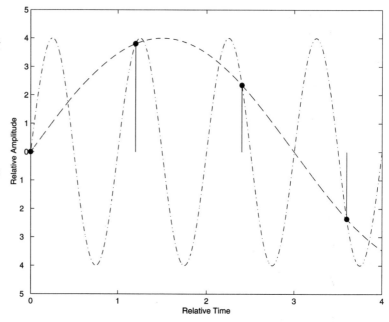

Figure 3-4. By rounding off the sample values to the nearest integer, as an example, the effect of quantization error in representing the waveform may be observed in a discrete time, discrete amplitude signal. The original signal is shown by the dashed line; the empty circles represent the quantized sample values; and the filled circles indicate the error for each sample point. The error signal is observed to lie within ±0.5 LSB.

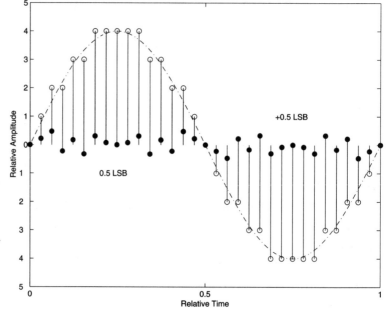

Digital Basics

As an example, the compact disc (CD) format uses 16-bit quantization, resulting in a technical SNR in the range of 96 dB. There are systems today that are technically capable of storing 24-bit samples that could, in theory, provide an even higher SNR and thus a lower amount of quantization error.

There is another method to reduce the amount of "audible" distortion in the quantization process. This is called *dithering*, which consists of adding a very low-level noise (of the order of an LSB) to the signal being quantized. Adding a suitable amount of dithering in an appropriately designed system can actually allow the system to attain precision beyond its stated specifications. This technique provides an audible improvement in the sound, but degrades the measured SNR, since noise is being added. This is particularly useful for low-level signals. A very simple explanation for why dithering works would be to say that the noise serves to break up any correlation between the signal and the quantization error—as the amplitude of a signal decreases, its quantization error gradually moves from a white noise spectrum to a correlated error.

The actual process of converting an analog signal to the digital domain is achieved with an Analog-to-Digital Converter (ADC). For proper operation, it is necessary to ensure that there is an anti-aliasing (low-pass) filter upstream for the ADC to limit the bandwidth; also, dithering may be employed in an attempt to linearize the system for low-level signals.

In summary, the process of converting an analog signal to its digital equivalent consists of taking a continuous time and continuous amplitude signal and converting it into a discrete time, discrete amplitude equivalent. In order to remain faithful to the original signal the sampling theorem must be followed. Due to the finite resolution of a digital system, there will always be quantization error. By increasing the number of bits or by adding dither, the quantization error may be reduced, but it cannot be eliminated.

Mathematical Transforms

Transforms are mathematical tools that allow movement from the time domain to the generalized frequency domain and vice versa. Some commonly used transforms in conjunction with analog signals are the Laplace and the Fourier transforms. These are called continuous transforms since they are applied to signals that are continuous in time and frequency.

The *Laplace transform* is a mathematical operation that maps a continuous time domain function $x(t)$ to the complex frequency domain $X(s)$. In this case, the complex frequency $s = s + jw$, that is, s has a real part s and an imaginary part w. This complex frequency lies in the two-dimensional s-plane having $Re(s)$ and $Im(s)$ as the axes. The Laplace variable, s, is commonly used to specify the transfer functions for analog circuits (analog filters, for example).

The Laplace transform is mathematically defined as:

$$X(s) = \int_{-\infty}^{\infty} x(t)e^{-st}dt \quad (3\text{-}1)$$

The inverse Laplace transform may be used to obtain the time domain signal $x(t)$ from the Laplace domain representation $X(s)$ by the following operation:

$$x(t) = \frac{1}{j2\pi}\int_{c-j\infty}^{c+j\infty} X(s)e^{st}ds \quad (3\text{-}2)$$

The *Fourier transform* is a special case of the Laplace transform. It may be used to determine the frequency response of an analog system. The Fourier transform is a mathematical operation that maps a continuous time domain function $x(t)$ to the frequency domain $X(jw)$. The Fourier transform may be obtained from the Laplace transform by setting $s = jw$,

Digital Basics

that is, by setting the real portion of s to zero. This means that the Fourier transform may be obtained from the imaginary axis of the Laplace plane. Therefore, the Fourier transform and the inverse Fourier transform are defined respectively as:

$$X(j\omega) = \int_{-\infty}^{\infty} x(t) e^{-j\omega t} dt \quad (3\text{-}3)$$

$$x(t) = \frac{1}{2\pi} \int_{-\infty}^{\infty} X(j\omega) e^{j\omega t} d\omega \quad (3\text{-}4)$$

Discrete transforms, on the other hand, are applied to signals that are discrete in time and frequency. The *Discrete Fourier Transform* (DFT), $X(m)$, is such a transform. It is the equivalent of a Fourier transform for a sampled signal $x(n)$. In theory, the discrete time signal is assumed to exist over all time; however, in practical situations only a finite number of samples are considered. This is called a finite Fourier transform. If the signal consists of N samples, then the N-point DFT and its inverse are respectively defined as:

$$X(m) = \sum_{n=0}^{N-1} x(n) W_N^{nm} \quad (3\text{-}5)$$

$$x(n) = \frac{1}{N} \sum_{m=0}^{N-1} X(m) W_N^{-nm} \quad (3\text{-}6)$$

where $W_N = e^{j2\pi/N}$. It should be noted that W_N^k is equal to W_N^{k+pN}, for any integer p. This shows that W_N has a period of N. As a result, $X(m) = X(m \pm pN)$ and $x(n) = x(n \pm pN)$. Thus, both $x(n)$ and $X(m)$ are periodic, and these sequences may be constructed for any n and m using a periodic extension from the fundamental period.

The DFT breaks down the time domain signal in terms of a harmonically related set of discrete frequencies, where the fundamental frequency

DSP Filters

is $2\pi F_s / N$. Examining the amplitude response at each of these discrete frequencies provides a measure of the spectrum of the signal. Hence, the DFT may be used to extract the frequency response of a digital system.

In implementing a DFT, the computational complexity is found to be of the order of N^2. There exists a collection of algorithms that better exploit the symmetric and periodic properties of the DFT called the *Fast Fourier transform* (FFT). The computational complexity is of the order of $N \log N$. The FFT algorithm typically requires N to be an integral power of two. Sequences of arbitrary length are zero padded to meet this requirement. Numerically, both the FFT and DFT produce the same result, but the FFT result is computed much faster. For $N = 1024$, the FFT represents a 100-fold increase in performance compared to the DFT. This may make the difference between an algorithm that may or may not run in real time.

The *z-transform*, $X(z)$, is equivalent to the Laplace transform for discrete sampled data and is the building block for digital filters. Hence, we would expect the DFT to be a special case of the z-transform, just as the Fourier transform is a special case of the Laplace transform in the analog domain. The z-transform is mathematically defined as:

$$X(z) = \sum_{n=-\infty}^{\infty} x(n) z^{-n} \quad (3\text{-}7)$$

Mathematically, it may be observed that substituting $z = e^{j 2\pi / N}$, results in the expression above equating to the definition of the DFT, Equation (3-5). This means that the values of N equally spaced points on the unit circle of the z-plane directly equals the N-point DFT.

Two important properties that the z-transform follows are:

1. The z-transform is linear — the z-transform of $ax(n) + by(n)$ is $aX(z) + bY(z)$.

2. The z-transform of $x(n-k)$, samples that have been delayed by k units, is given by $z^k X(z)$.

Digital Basics

The second property shows that the expression z^{-1} in the z-domain is equivalent to a unit delay in the discrete time domain. It is thus called the unit delay element. This property could be used to find the z-transform of a time domain equation directly. In discrete systems, the input and output only exist at sampled times, thus these systems are described in terms of difference equations.

Using the linearity and time delay property from above, it is possible to take the z-transform of the following difference equation:

$$y[n] = x[n] - 0.3x[n-1] + 0.2y[n-1] \quad (3\text{-}8)$$

Therefore, the z-transform would be:

$$Y(z) = X(z) - 0.3z^{-1}X(z) + 0.2z^{-1}Y(z) \quad (3\text{-}9)$$

Rearranging and grouping X(z) and Y(z) terms, yields:

$$Y(z)[1 - 0.2z^{-1}] = X(z)[1 - 0.3z^{-1}] \quad (3\text{-}10)$$

The *transfer function*, H(z)=Y(z)/X(z), is defined by the ratio of the output to the input transforms:

$$H(z) = \frac{Y(z)}{X(z)} = \frac{1 - 0.3z^{-1}}{1 - 0.2z^{-1}} \quad (3\text{-}11)$$

Digital Filter

Consider a general difference equation of the form:

$$\begin{aligned}y[n] + b_1 y[n-1] + b_2 y[n-2] + \cdots + b_N y[n-N] = \\ a_0 x[n] + a_1 x[n-1] + a_2 x[n-2] + \cdots + a_M x[n-M]\end{aligned} \quad (3\text{-}12)$$

where a_i and b_j are coefficients, x[n-i] represents the ith previous input sample and y[n-j] represents the jth previous output sample. Thus, x[n] and y[n] represent the present input and output sample, respectively. This discrete time equation defines a digital filter.

27

DSP Filters

The equation above may be rearranged and rewritten in mathematical shorthand as:

$$y[n] = \sum_{i=0}^{M} a_i x[n-i] - \sum_{j=1}^{N} b_j y[n-j] \quad (3\text{-}13)$$

Taking the z-transform of this equation yields:

$$Y(z) = \sum_{i=0}^{M} a_i z^{-i} X(z) - \sum_{j=1}^{N} b_j z^{-j} Y(z) \quad (3\text{-}14)$$

$$Y(z) = X(z) \sum_{i=0}^{M} a_i z^{-i} - Y(z) \sum_{j=1}^{N} b_j z^{-j} \quad (3\text{-}15)$$

Regrouping the equation above yields:

$$Y(z)[1 + \sum_{j=1}^{N} b_j z^{-j}] = X(z) \sum_{i=0}^{M} a_i z^{-i} \quad (3\text{-}16)$$

The transfer function, H(z), now becomes:

$$H(z) = \frac{Y(z)}{X(z)} = \frac{\sum_{i=0}^{M} a_i z^{-i}}{1 + \sum_{j=1}^{N} b_j z^{-j}} = \frac{N(z)}{D(z)} \quad (3\text{-}17)$$

Mathematically this shows that the transfer function of a digital filter is completely determined by the value of the coefficients being used. In other words, this implies that by manipulating the values of the coefficients it is possible to change the characteristics of the transfer function that the input signal is subjected to. This fact may be considered the basis for the flexibility that is found in digital filter implementations.

Digital Basics

Equation (3-17) is the ratio of two polynomials in z^{-1}, consisting of the numerator term $N(z)$ and the denominator term $D(z)$. The numerator consists of M roots, called the zeros of the system since the transfer function is equal to zero at these values. Similarly, the denominator consists of N roots, called the poles of the system since the denominator becomes zero at these values and the transfer function is not defined.

The location of the poles and zeros in the complex z-plane provides an intuitive feel for the transfer function as follows. If one were to place long, thin poles at the pole locations and then stretch a thin membrane over the z-plane such that it touched the z-plane at the zero locations, then the membrane reflects the magnitude response of the transfer function. In particular, if one were to trace the contour of the membrane along the unit circle (starting at $z = 0$ and moving counterclockwise until $z = 1$), one would obtain the traditional frequency response of the digital filter.

Examination of the numerator and denominator terms in Equation (3-17) shows that the zeros of the system are contributed by the a_i coefficients, which are associated with the present and past input samples. The poles of the system are contributed by the b_j coeffifcients, which are associated with the past output samples.

Unit Circle

When designing a digital filter, the stability of the filter is something that should be given due consideration. An unstable filter is certainly not desirable since it could go into oscillation or produce an output that increases in an "unbounded" manner until it reaches either positive or negative saturation—this is not the kind of filtering one normally employs in audio systems, unless this is the effect desired.

The condition for testing stability of a digital filter is relatively straightforward. The poles of the digital filter need to lie inside the unit circle of

DSP Filters

the z-plane. This may be done either by plotting the locations of the individual poles and graphically ensuring that they do indeed lie inside the unit circle or by ensuring that the magnitude of the complex value of the pole location is less than one (that is, inside the unit circle).

If we move into the practical world, there are two things with regard to pole locations that we need to be aware of:

1. The amount of precision available to represent the coefficients in a digital signal processor is finite—usually the word length is used. In some cases, double or extended precision techniques may be employed to get greater precision—even in these cases, there is no getting away from the fact that there is a finite limit to the precision. Finite precision may have an impact on the real implementation characteristics of the digital filter because it has the potential of altering the locations of the poles and zeros. In extreme cases, this may cause a theoretical pole that lay *inside* the unit circle to move *outside*, which results in instability.

2. A technique that is sometimes used by engineers when transitioning from one filter configuration to another is to interpolate the intermediate values of the coefficient over a number of audio samples. Although the initial and final filter configurations may be stable ones (that is, have their poles inside the unit circle), some of the intermediate stages *may* have poles lying outside the unit circle— once again resulting in instability.

Digital Signal Processing Operations

$$y[n] = a_0 x[n] + a_1 x[n-1] + a_2 x[n-2] + \cdots + \quad (3\text{-}18)$$
$$a_M x[n-M] - b_1 y[n-1] - b_2 y[n-2] + \cdots - b_N y[n-N]$$

The difference equation shown in Equation (3-18) may be obtained by rearranging Equation (3-12) or expanding Equation (3-13). The ex-

Digital Basics

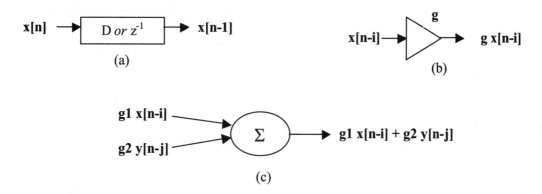

Figure 3-5. Hardware realizations of the operations required to implement a digital filter (a) Delay element, (b) Multiplication operation, and (c) Summation operation.

pression shows how the present output y[n] may be calculated from past and present input values and past output values.

Thus, the present output is the summation of a weighted combination of past inputs and outputs. This may be realized by using three operations:

1. *Delay* element to produce delayed versions of the input (x[n-i]) or output (y[n-j]). A unit delay element would produce an output of x[n-1] given an input x[n]. It is commonly represented as a box with a z^1 or D, as shown in *Figure 3-5a*. A number of these elements may be cascaded to achieve longer delays. Delay elements may be implemented using shift registers in hardware or they may be memory locations that are appropriately accessed by the processor. Unlike the analog domain, the signal integrity is completely maintained during the delay process in the digital domain.

2. *Multiplication* to allow the delayed input and output signals to be weighted appropriately by coefficients a_i and b_j, respectively. If a multiplier has a gain of g, then for an input of x[n-i], the output would be g x[n-i], as shown in *Figure 3-5b*. Thus, the first two elements allow the creation of terms such as a_i x[n-i] and b_j y[n-j].

31

DSP Filters

3. *Summation* to combine the weighted delayed versions of the input and output to produce the present output. This is shown in *Figure 3-5c*. This allows the partial terms to be combined to form the final output.

Using these three operations, any digital filter may be conceptually implemented from a hardware perspective. Unlike analog filters, digital filters are mostly implemented in software using a digital signal processor (DSP). In this case, the delay element consists of implementing buffers using memory locations. The multiplication and summation elements are usually combined in a single multiply-accumulate instruction during which two operands are multiplied and added to a running sum. This results in a more efficient filter implementation in software. Specialized DSP chips provide specialized hardware to facilitate implementing digital filter operations, since it is so fundamental to digital signal processing.

Types of Digital Filters

There are two basic types of digital filters:

(a) Infinite Impulse Response (IIR) or recursive filter, and

(b) Finite Impulse Response (FIR) or nonrecursive filter.

Both of these may be derived from the generic digital filter Equation (3-13) that is reproduced here again for the sake of completeness.

$$y[n] = \sum_{i=0}^{M} a_i x[n-i] - \sum_{j=1}^{N} b_j y[n-j] \quad (3\text{-}19)$$

In *Figure 3-6*, the hardware realization for this equation is shown. Since the output values are delayed and then fed back, this is termed a recursive or IIR filter. Owing to the feedback paths that are present, this filter has two potential problems:

Digital Basics

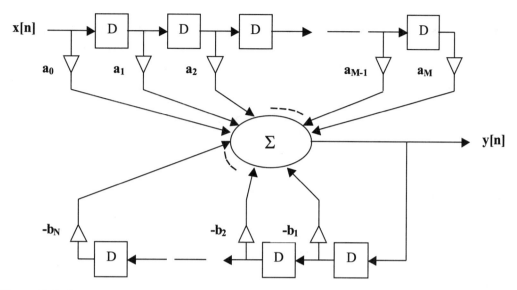

Figure 3-6. The realization of a generic recursive or IIR digital filter.

1. It may be unstable.
2. It may demonstrate nonlinear phase response.

Care needs to be taken in designing them for audio applications.

If the assumption is made that all the coefficients b_j are equal to zero, then the resulting equation is that of an FIR filter:

$$y[n] = \sum_{i=0}^{M} a_i x[n-i] \quad (3\text{-}20)$$

Hardware realization of this difference equation using the basic building blocks only requires feed-forward terms as shown in Figure 3-7. These filters are potentially stable and display linear phase response.

Looked at from the viewpoint of the transfer function definition in terms of poles and zeros of the system, it may be observed that the FIR

DSP Filters

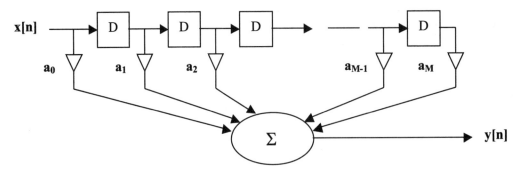

Figure 3-7. The realization of a generic nonrecursive or FIR digital filter.

filter has no poles. The lack of poles further reinforces the fact that an FIR filter is stable.

Comparison of IIR and FIR Filters

As just seen, if the transfer function of a filter is the ratio of two polynomials in z^1 then the underlying filter is a recursive or IIR filter. However, if the denominator term of the transfer function is unity then the transfer function will be a polynomial in z^1, and the underlying filter will be a nonrecursive or FIR filter.

Owing to the lack of any feedback components, FIR filters are inherently stable since there are no poles in the transfer function (where the response could potentially blow up). From an analytical standpoint, a FIR implementation is easier to understand. The feedback components in an IIR design could make an IIR implementation more difficult to follow from an intuitive standpoint.

Introduction of the feedback paths allows an IIR implementation to have higher selectivity and sharper cutoff characteristics compared to a FIR filter of the same order. This makes a FIR filter more expensive to implement in terms of computation time and memory requirements.

Digital Basics

FIR filters exhibit linear phase characteristics. IIR filters, on the other hand, exhibit phase nonlinearity due to the feedback paths. They could be designed to have linear phase characteristics using phase equalization.

The Biquad Section

From the properties just described, the preference in audio would be to use the stable, linear-phase FIR filters. However, the overriding expense of such an implementation makes it impractical for most applications. Hence, the majority of audio processing in the spectral domain (such as tone control, graphic, or parametric equalization) tends to use IIR filters. The basic building block in most cases is often a biquadratic section, also known as a *biquad*. This is analogous to a second-order filter section in the analog domain.

The biquad section may be defined by the following generic digital filter equation:

$$y[n] = a_0 x[n] + a_1 x[n-1] + a_2 x[n-2] - b_1 y[n-1] - b_2 y[n-2] \quad (3\text{-}21)$$

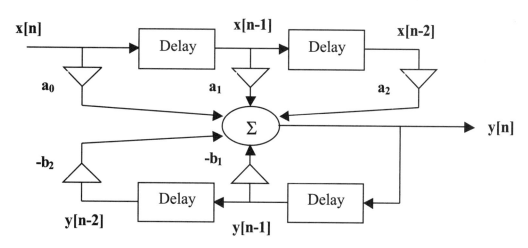

Figure 3-8: The realization of a biquadratic section.

DSP Filters

Figure 3-8 shows the realization of the biquad section. It utilizes the present input sample, *x[n]*; two past input samples, *x[n-1]* and *x[n-2]*; two past output samples, *y[n-1]* and *y[n-2]*; and five associated coefficients, a_0, a_1, a_2, $-b_1$, and $-b_2$, to generate the present output sample, *y[n]*. These coefficients are the key elements that provide the biquad with a lot of flexibility. By manipulating the values of the coefficients, the nature of the filter can be changed completely as the subsequent chapters will demonstrate. The beauty of digital systems being implemented in a general-purpose programmable digital signal processor is that it is relatively easy to manipulate the coefficient values. It is also possible to combine these biquad filter sections in a cascade or parallel fashion to create more complicated filtering structures. In essence, this implies that the biquad section is a basic building block for filtering applications in the digital domain.

Returning Back to the Analog Domain

As was mentioned before, the world around us is analog. Hence, after the audio signals have been subjected to signal processing in the digital domain, they must be returned to the analog domain so that human beings are able to hear them. This task is generally assigned to a chip called a Digital-to-Analog converter (DAC). Thus the DAC serves the opposite function of the ADC. Just as the ADC determines the overall fidelity of the incoming analog signal, the DAC determines the final quality of the digital signal when it is converted to the analog domain for listening purposes.

Just as there is a low-pass filter before an ADC to minimize aliasing effects, similarly, there is a low-pass filter after the DAC. Its purpose is to reconstruct a smooth analog output waveform from the staircase-like samples put out by the DAC at each sampling period.

Summary

A digital system, when properly designed, can provide a lot of flexibility and a myriad of signal processing tools, making it a versatile platform for manipulating signals compared to analog systems. In order to harness this, it is necessary to sample the analog signal and digitize it by using analog-to-digital converters. Care needs to be taken to ensure that aliasing is prevented and dithering techniques are employed when necessary to minimize the effects of quantization error.

Digital filters form the heart of signal processing tools in the time domain. By manipulating the coefficients that form a digital filter, it is possible to control the nature and behavior of the signals being processed. This allows a real-time system to update coefficients on the fly and change the characteristics of the processing block in a "soft" manner. In particular, the biquadratic section, with five controlling coefficients, may be viewed as a basic building block for digital filters.

Being analog creatures, the digital signals need to be returned back to the analog domain using digital-to-analog converters. Once a signal is digitized, it should undergo all necessary processing in the digital domain before being converted back to the analog domain. Reducing the number of ADC and DAC stages could result in a cleaner signal chain.

4

First-Order Low-Pass Filter

Analog Filter Network

A low-pass filter can be implemented using the RC network shown in *Figure 4-1*. Note that a combination of R and L can also lead to low-pass filter characteristics, but it will have a different expression for the cutoff frequency Ω_c.

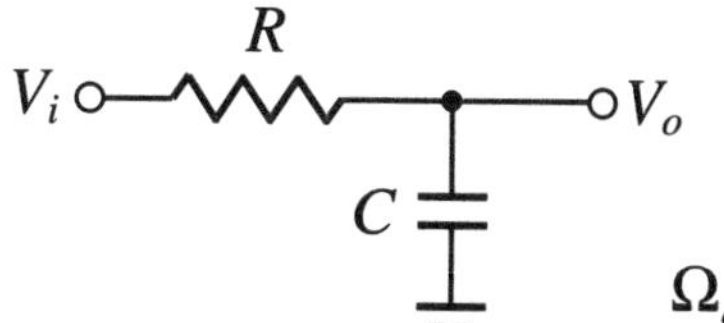

Figure 4-1. RC low-pass filter network

$$\Omega_c \equiv \frac{1}{RC}$$

The impedances associated with any ideal resistor R and capacitor C are:

$$Z_R = R \qquad Z_C = \frac{1}{sC} \qquad (4\text{-}1)$$

DSP Filters

where the impedances Z are represented by complex numbers, since the Laplace variable s is complex. When performing calculations of complex impedance or evaluation of a filter transfer function, it is assumed that $s = j\Omega$, where $j \equiv \sqrt{-1}$ and $\Omega \equiv 2\pi f$ (f is frequency in Hertz). Note that it is essential to work with a consistent set of units of measure for R, C, and f — the simplest of which is *ohms, farads,* and *Hertz*.

The filter of *Figure 4-1* is expressed as the ratio of the output V_o to input V_i and is equivalent to the ratio of the capacitive impedance to the total impedance of the network:

$$\frac{V_o}{V_i} = \frac{Z_C}{Z_C + Z_R} \quad (4\text{-}2a)$$

Combining the expressions for complex impedance from Equations (4-1) for R and C results in a filter *transfer function* $H(s)$:

$$H(s) \equiv \frac{V_o}{V_i} = \frac{1/(sC)}{1/(sC) + R} \quad (4\text{-}2b)$$

where $H(s)$ is defined as the ratio of V_o to input V_i. $H(s)$ can be manipulated by dividing both the numerator and denominator by R and multiplying by s:

$$H(s) = \frac{\frac{1}{RC}}{s + \frac{1}{RC}} \quad (4\text{-}2c)$$

The result is a simplified representation of the complex low-pass filter transfer function:

$$H(s) = \frac{\Omega_c}{s + \Omega_c} \quad (4\text{-}3)$$

where,

$$\Omega_c \equiv \frac{1}{RC} \quad (4\text{-}4)$$

First-Order Low-Pass Filter

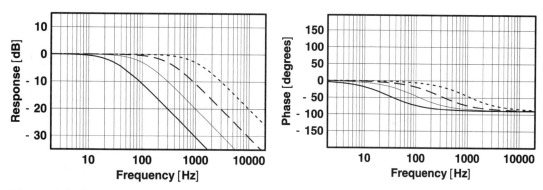

Figure 4-2. Gain response and phase plots for an analog low-pass filter: solid line, $f_c = 30$; thin solid line, $f_c = 100$; dashed line, $f_c = 300$; dotted line, $f_c = 1000$

The filter transfer function of Equation (4-3) is a precise description of the first-order low-pass filter response. However, since $H(s)$ is complex, it can be an inconvenient description. A more useful set of formulas for the filter frequency response is the magnitude $G(f)$ and the phase angle $\phi(f)$ of $H(s)$:

$$G(f) = \frac{f_c}{\sqrt{f^2 + f_c^2}} \qquad \phi(f) = 180\left[\frac{1}{\pi}\tan^{-1}\left(\frac{-f}{f_c}\right)\right] \qquad (4\text{-}5a)$$

$$f_c \equiv \frac{1}{2\pi RC} \qquad (4\text{-}5b)$$

Magnitude and phase response of the analog first-order low-pass filter

Figure 4-2 shows several examples of gain and phase response curves for the analog first-order low-pass filter using several values of cutoff frequency: thick solid line, $f_c = 30$ Hz; thin solid line, $f_c = 100$ Hz; dashed line, $f_c = 300$ Hz; and dotted line, $f_c = 1000$ Hz. The magnitude of the gain and phase at the cutoff frequency can easily be determined by substituting $f = f_c$ in Equations (4-5a) and (4-5b):

DSP Filters

$$G(f_c) = \frac{f_c}{\sqrt{f_c^2 + f_c^2}} = \frac{1}{\sqrt{2}} \qquad (4\text{-}6a)$$

$$\phi(f_c) = 180\left[\frac{1}{\pi}\tan^{-1}\left(-\frac{f_c}{f_c}\right)\right] = -45° \qquad (4\text{-}6b)$$

Gain and phase response of analog first-order low-pass filter at its cutoff frequency

Digital Filter Network

A first-order digital low-pass filter is constructed by implementing a *running average* on a stream of sampled data, using one previous input and one previous output for each new input and output value. Because of the use of previous outputs, this type of filter is called a *recursive* filter or an *infinite impulse response* (IIR) filter. An impulse response is the characteristic filter output caused by an input composed of a single pulse (a 1 followed by all 0s). The impulse response is infinite in duration, since in theory, the output never decays to absolute zero. However, since a real filter is based on finite numbers stored in memory in a microprocessor, DSP, or computer system, an actual IIR filter impulse response is finite in length, or it may oscillate indefinitely around some small value — a phenomenon known as *limit cycling*.

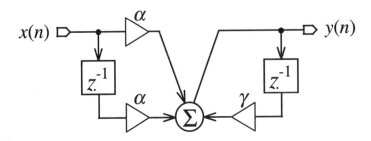

Figure 4-3. Digital low-pass filter network

First-Order Low-Pass Filter

The basic components of the digital filter network, as shown in *Figure 4-3*, are the *multiplier* (triangle), *adder* (circle), and *delay element* (rectangle). In a purely hardware implementation, the identification of these components is important, but in a purely software implementation, the precise classification of these components is not so important. When using a general-purpose programmable digital signal processor implementation, these components correspond to key instructions of the DSP's *instruction set*. Typically, in a pure hardware or DSP implementation, the multiplier and adder are combined into a single unit called the *accumulator*.

Like the s-domain transfer function of Equation (4-3), the transfer function of the low-pass network of *Figure 4-3* describes its response characteristics, but instead is a function of the z-transform:

$$H(z) = \frac{\alpha \left(1 + z^{-1}\right)}{1 - \gamma z^{-1}} \quad (4\text{-}7)$$

Recall that the s variable of the Laplace transform (or s-transform) represents a point in the complex plane such that the analog filter response characteristics are found along the vertical line determined by setting $s = j2\pi f$. In a similar manner, the z variable of the z-transform is again a point in the complex plane, but the digital filter response is found along the unit circle at $z = e^{j2\pi f / f_s}$, where f_s is the sample frequency.

The coefficients α and γ in Equation (4-7) are directly related to the filter cutoff frequency f_c, by:

$$\beta = \frac{1}{2} \frac{1 - \sin(2\pi f_c / f_s)}{1 + \sin(2\pi f_c / f_s)}$$

$$\gamma = \left(\tfrac{1}{2} + \beta\right)\cos(2\pi f_c / f_s) \quad (4\text{-}8)$$

$$\alpha = (1 - \gamma)/2$$

43

DSP Filters

The filter transfer function of Equation (4-7) is a precise description of the digital low-pass filter response. However, since $H(z)$ is complex, it can be an inconvenient description, as in the case of $H(s)$. A more useful set of formulas for the filter frequency response is the magnitude $G(f)$ and the phase angle $\phi(f)$ of $H(z)$:

$$G(f) = \sqrt{\frac{(1+\cos\theta)(1-\cos\theta_c)}{2(1-\cos\theta\cos\theta_c)}} \qquad (4\text{-}9a)$$

$$\phi(f) = 180\left[\frac{1}{\pi}\tan^{-1}\left(\frac{-(1+\cos\theta_c)\sin\theta}{(1+\cos\theta)\sin\theta_c}\right)\right] \qquad (4\text{-}9b)$$

where, $\theta \equiv 2\pi f / f_s \qquad \theta_c \equiv 2\pi f_c / f_s$

Magnitude and phase response of the digital first-order low-pass filter

Figure 4-4 shows several examples of gain and phase response curves for the first order IIR low-pass filter using several values of cutoff frequency, with f_s = 11025 Hz: thick solid line, f_c = 30 Hz; thin solid line, f_c = 100 Hz; dashed line, f_c = 300 Hz; and dotted line, f_c = 1000 Hz. Note the differences between the response of the analog filter of Figure 4-2 and the digital filter of Figure 4-4. The digital filter response goes to zero

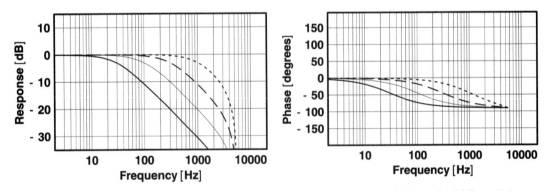

Figure 4-4. Gain and phase of a first-order digital low-pass filter with f_s = 11025: solid line, f_c = 30; thin solid line, f_c = 100; dashed line, f_c = 300; and dotted line, f_c = 1000

First-Order Low-Pass Filter

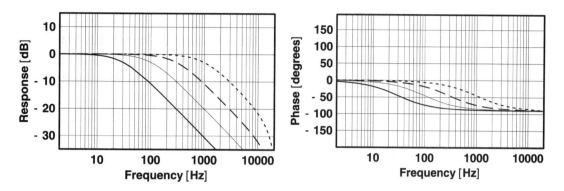

Figure 4-5. Gain and phase of a first-order digital low-pass filter with $f_s = 44100$ Hz: solid line, $f_c = 30$; thin solid line, $f_c = 100$; dashed line, $f_c = 300$; and dotted line, fc = 1000

(minus infinity on a dB plot) at the *Nyquist* frequency $f_n = f_s/2$, where in this example, $f_s = 11025$ Hz and $f_n = 5512.5$ Hz. If the sample frequency is set to $f_s = 44100$ Hz, as shown in *Figure 4-5*, the digital frequency response plots will appear nearly identical to those of *Figure 4-2*.

This can be demonstrated by recalling that the trigonometric functions, cos x and sin x, can be represented by:

$$\left.\begin{array}{l}\cos\theta \to 1-\tfrac{1}{2}\theta^2 \\ \sin\theta \to \theta\end{array}\right\} \quad \text{for } \theta \to 0 \quad (4\text{-}10)$$

and then substituting in Equation (4-9a):

$$G(f) \to \sqrt{\frac{\left(1+1-\tfrac{1}{2}\theta^2\right)\left(1-1+\tfrac{1}{2}\theta_c^2\right)}{2\left[1-\left(1-\tfrac{1}{2}\theta^2\right)\left(1-\tfrac{1}{2}\theta_c^2\right)\right]}} \quad \right\} \quad \text{for } \theta \to 0$$

$$= \sqrt{\frac{2\left(\tfrac{1}{2}\theta_c^2\right)}{2\left[1-\left(1-\tfrac{1}{2}\theta_c^2-\tfrac{1}{2}\theta^2+\ldots\right)\right]}}$$

$$\approx \frac{\theta_c}{\sqrt{\theta^2+\theta_c^2}} = \boxed{\frac{f_c}{\sqrt{f^2+f_c^2}}}$$

(4-11)

DSP Filters

Comparing the result of Equation (4-11) with the analog gain response of Equation (4-5a), it can be seen that they are equivalent. A similar result is obtained by substituting Equation (4-10) in the IIR phase formula of Equation (4-9b). It will be identical to the analog low-pass phase formula of Equation (4-5a).

Difference Equation

In order to implement the digital IIR filter a *difference equation* is used. As previously mentioned, the IIR filter implementation is simply a running average of the input and output data:

$$y(n) = \alpha \left[x(n) + x(n-1)\right] + \gamma\, y(n-1) \quad (4\text{-}12)$$

where $x(n)$ is the current input sample; $x(n\text{-}1)$ is the previous input; $y(n\text{-}1)$ is the previous output; and $y(n)$ is the current output. The filter coefficients α and γ are defined in Equation (4-8), as well as in (4-13b). Note that the coefficient β has been eliminated in Equation (4-13b) by direct substitution into g of Equation (4-8). Also note that the notation used in Chapter 3 for time domain samples is equivalent, i.e., $x[n] = x(n)$.

There are several ways to interpret the meaning as well as application of the low-pass filter difference equation above:

- Equation (4-12) is simply the result of applying the inverse z-transform on the z-transfer function $H(z)$ from Equation (4-7).

$$y(n) = \alpha \left[x(n) + x(n-1)\right] + \gamma\, y(n-1) \quad (4\text{-}13a)$$

$$\gamma = \frac{\cos\theta_c}{1 + \sin\theta_c} \qquad \alpha = (1-\gamma)/2 \quad (4\text{-}13b)$$

$$\text{where,}\ \theta \equiv 2\pi f / f_s \qquad \theta_c \equiv 2\pi f_c / f_s$$

Difference equation of the first-order IIR low-pass filter with coefficient formulas

First-Order Low-Pass Filter

- The input is a large buffer of data, such as that from a digital recording. For example, a digital telephone answering machine would store 80,000 samples for a 10-second message, assuming a sample frequency of 8000 Hz. In that case, the sample index n of Equation (4-12) is just the index of the input buffer (let n start at 0 and go to 79999). Each output $y(n)$ is computed by evaluating Equation (4-12) for every value of n from 0 to 79999. Note that $x(-1) = y(-1) = 0$ is the initial condition.
- The input is a data stream where at every sample time $T = 1/f_s$, an input sample $x(n)$ is available for processing via a clocked system such as an interrupt. In this case, two state variables $x1$ and $y1$ must be evaluated after each computation of $y(n)$.

The following C code implements the first-order low-pass difference equation:

```
//
// cook_lowpass1.cpp
//
// Implementation of a simple 1st Order Low-pass filter Stage,
// class Clowpass1FilterStage

#include "cook.h"

Clowpass1FilterStage::Clowpass1FilterStage()
{
        x1 = 0;
        y1 = 0;
}

Clowpass1FilterStage::~Clowpass1FilterStage()
{
}

void Clowpass1FilterStage::execute_filter_stage()
{
        y = alpha * (x + x1) + gamma * y1 ;
        x1 = x;
        y1 = y;
        y = y * gain;
}
```

5
First-Order
High-Pass Filter

Analog Filter Network

A high-pass filter can be implemented using the RC network shown in *Figure 5-1*. Similar to the low-pass case of Chapter 4, a combination of *R* and *L* may also lead to high-pass filter characteristics, but will have a different expression for the cutoff frequency Ω_c.

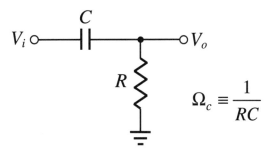

Figure 5-1. RC high-pass filter network

DSP Filters

Repeated below are the R and C impedance formulas from the previous chapter:

$$Z_R = R \qquad Z_C = \frac{1}{sC} \qquad (5\text{-}1)$$

where $s = j\Omega = 2\pi j f$, R has units of *ohms*, C is in *farads*, and f is described by *Hertz*. The impedance values in Equation (5-1) are in general complex values, since the Laplace variable s is complex (recall that $j \equiv \sqrt{-1}$ and $\Omega \equiv 2\pi f$).

The RC filter of *Figure 5-1* can be described as the ratio of the output V_o to input V_i, and is equivalent to the ratio of the resistive impedance to the total impedance of the network (this is essentially a complex *voltage divider*):

$$\frac{V_o}{V_i} = \frac{Z_R}{Z_C + Z_R} \qquad (5\text{-}2a)$$

Combining the expressions for complex impedance from Equations (5-1) for R and C results in a high-pass filter *transfer function H(s)*:

$$H(s) \equiv \frac{V_o}{V_i} = \frac{R}{1/(sC) + R} \qquad (5\text{-}2b)$$

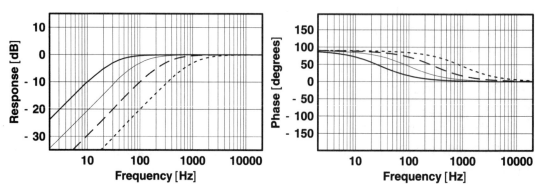

Figure 5-2. Gain response and phase plots for analog high-pass filter: solid line, f_c = 30; thin solid line, f_c = 100; dashed line, f_c = 300; and dotted line, f_c = 1000

First-Order High-Pass Filter

where $H(s)$ is defined as the ratio of V_o to input V_i. $H(s)$ can be manipulated by dividing both the numerator and denominator by R and multiplying by s:

$$H(s) = \frac{s}{s + \frac{1}{RC}} \qquad (5\text{-}2c)$$

The result is a simplified representation of the complex high-pass filter transfer function:

$$H(s) = \frac{s}{s + \Omega_c} \qquad (5\text{-}3)$$

where,

$$\Omega_c \equiv \frac{1}{RC} \qquad (5\text{-}4)$$

The complex filter transfer function of Equation (5-3) contains all of the information needed to describe the frequency response of the first-order high-pass filter of *Figure 5-1*. This single complex function $H(s)$ can be equivalently expressed as two real functions representing the magnitude response $G(f)$ and the phase response $\phi(f)$:

$$G(f) = \frac{f}{\sqrt{f^2 + f_c^2}} \qquad \phi(f) = 180\left[\frac{1}{\pi}\tan^{-1}\left(\frac{f_c}{f}\right)\right] \qquad (5\text{-}5a)$$

$$f_c \equiv \frac{1}{2\pi RC} \qquad (5\text{-}5b)$$

Magnitude and phase response of the analog first-order high-pass filter

Figure 5-2 shows several examples of magnitude and phase response curves for the analog first-order high-pass filter using several values of cutoff frequency: thick solid line, f_c = 30 Hz; thin solid line, f_c = 100 Hz; dashed line, f_c = 300 Hz; and dotted line, f_c = 1000 Hz.

DSP Filters

The magnitude and phase response at the cutoff frequency is found by substituting $f = f_c$ in Equations (5-5a):

$$G(f_c) = \frac{f_c}{\sqrt{f_c^2 + f_c^2}} = \frac{1}{\sqrt{2}} \qquad (5\text{-}6a)$$

$$\phi(f_c) = 180\left[\frac{1}{\pi}\tan^{-1}\left(\frac{f_c}{f_c}\right)\right] = 45° \qquad (5\text{-}6b)$$

Gain and phase response of analog first-order high-pass filter at its cutoff frequency

Digital Filter Network

As described in the previous chapter, a first-order digital high-pass filter can be constructed by implementing a *running average* on a stream of sampled data, using one previous input and one output for each new input sample. Due to a single input pulse, the filter output is infinite in duration and never decays to zero when implemented in the hypothetical case of infinite precision arithmetic. However, since a practical digital filter is implemented using finite precision numbers (16-bit, 24-bit, 32-bit, etc.), an IIR filter impulse response can end up being finite in length. After a relatively long time, which may only be milliseconds, the impulse response may decay to a very small number or oscillate indefinitely around some small error value.

As shown in *Figure 5-3*, the basic components of the digital filter network are the *multiplier* (triangle), *adder* (circle), and *delay element* (rect-

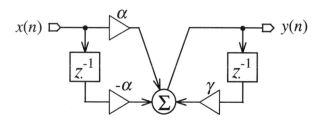

Figure 5-3. Digital high-pass filter network

First-Order High-Pass Filter

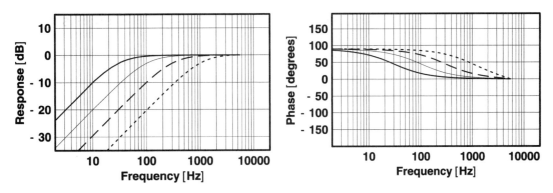

Figure 5-4. Gain and phase of first-order digital high-pass filter with f_s = 11025: solid line, f_c = 30; thin solid line, f_c = 100; dashed line, f_c = 300; dotted line, f_c = 1000

angle). When using a general-purpose digital signal processor (DSP), these components correspond to key instructions of the DSP's *instruction set*. Typically, in a purely hardware or DSP implementation, a multiplier and adder are combined into a new separate function called the *accumulator*. In a purely software implementation (see Project chapters 16-21), these key filter components are simply the conventional *add* and *multiply* functions. The *delay* function is implemented in software by equating program variables, as shown in the next section.

Recall that the *s* variable of the s-transform (or Laplace transform) represents a point in the complex plane such that the analog filter response characteristics are found along the vertical line determined by setting $s = j2\pi f$. In a similar manner, the *z* variable of the z-transform is again a point in the complex plane, but now the digital filter response is found along the unit circle at $z = e^{j2\pi f / f_s}$, where f_s is the sample frequency.

As in the case of the s-domain transfer function described by Equation (5-3), the transfer function $H(z)$ of the high-pass network of Figure 5-3 describes its frequency response characteristics:

$$H(z) = \frac{\alpha \left(1 - z^{-1}\right)}{1 - \gamma \, z^{-1}} \qquad (5\text{-}7)$$

DSP Filters

The coefficients α and γ in Equation (5-7) are directly related to the filter cutoff frequency f_c, by:

$$\beta = \frac{1}{2}\frac{1-\sin(2\pi f_c / f_s)}{1+\sin(2\pi f_c / f_s)}$$

$$\gamma = \left(\tfrac{1}{2}+\beta\right)\cos(2\pi f_c / f_s) \qquad (5\text{-}8)$$

$$\alpha = (1+\gamma)/2$$

Equation (5-7) is a concise and exact description of the first-order, digital high-pass filter's frequency response. However, since $H(z)$ is complex, it may not be the most convenient description. An equivalent mathematical representation of the digital frequency response where magnitude response $G(f)$ and phase response $\phi(f)$ are given by the following set of real formulas:

$$G(f) = \sqrt{\frac{(1-\cos\theta)(1+\cos\theta_c)}{2(1-\cos\theta \cos\theta_c)}} \qquad (5\text{-}9a)$$

$$\phi(f) = \frac{180}{\pi}\tan^{-1}\frac{(1-\cos\theta_c)\sin\theta}{(1-\cos\theta)\sin\theta_c} \qquad (5\text{-}9b)$$

where, $\theta \equiv 2\pi f / f_s \qquad \theta_c \equiv 2\pi f_c / f_s$

Magnitude and phase response of the IIR first-order high-pass filter

Figure 5-4 shows several examples of magnitude and phase response curves for the first-order IIR high-pass filter, using several values of cutoff frequency, with f_s = 11025 Hz: thick solid line, f_c = 30 Hz; thin solid line, f_c = 100 Hz; dashed line, f_c = 300 Hz; and dotted line, f_c = 1000 Hz. Similarly, Figure 5-5 shows the equivalent response curves at a higher sample frequency of f_s = 44100 Hz.

First-Order High-Pass Filter

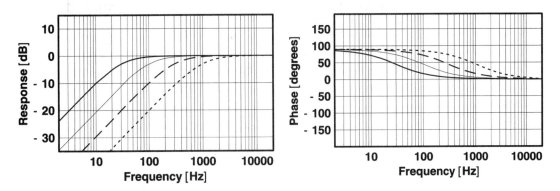

Figure 5-5. Gain and phase of first-order digital high-pass filter with fs = 44100 Hz: solid line, fc = 30; thin solid line, fc = 100; dashed line, fc = 300; dotted line, fc = 1000

Note that when $f_c \ll f_n$ or $f \ll f_n$ (where f_n is the Nyquist frequency, $f_n = \tfrac{1}{2} f_s$), the digital frequency response approaches that of the analog frequency response. This can be demonstrated by recalling that the trigonometric functions, cos x and sin x, can be represented by:

$$\left. \begin{array}{l} \cos\theta \to 1 - \tfrac{1}{2}\theta^2 \\ \sin\theta \to \theta \end{array} \right\} \text{ for } \theta \to 0 \quad (5\text{-}10)$$

Substituting Equations (5-10) into Equation (5-9a), results in:

$$G(f) \to \sqrt{\frac{\left(1 - 1 + \tfrac{1}{2}\theta^2\right)\left(1 + 1 - \tfrac{1}{2}\theta_c^2\right)}{2\left[1 - \left(1 - \tfrac{1}{2}\theta^2\right)\left(1 - \tfrac{1}{2}\theta_c^2\right)\right]}} \quad \text{for } \theta \to 0$$

$$= \sqrt{\frac{2\left(\tfrac{1}{2}\theta^2\right)}{2\left[1 - \left(1 - \tfrac{1}{2}\theta_c^2 - \tfrac{1}{2}\theta^2 + ...\right)\right]}} \quad (5\text{-}11)$$

$$\approx \frac{\theta}{\sqrt{\theta^2 + \theta_c^2}} = \frac{f}{\sqrt{f^2 + f_c^2}}$$

By comparing the result in Equation (5-11) with the analog magnitude response of Equation (5-5a), it can be seen that they are equiva-

DSP Filters

lent. A similar result is obtained by substituting Equation (5-10) in the IIR phase response formula of Equation (5-9b): it will be identical to the analog high-pass phase response formula of Equation (5-5a). This behavior can be summarized by the following principle, previously discussed in Chapter 4:

> As the sample frequency f_s of an IIR filter is increased, the frequency response approaches that of its analog counterpart. Likewise, as the cutoff frequency f_c and operating frequency f is decreased, the digital filter frequency response approaches that of its analog filter counterpart.

Difference Equation

The digital IIR filter is implemented in software using a *difference equation*, a running recursive average of the input and output data:

$$y(n) = \alpha\left[x(n) - x(n-1)\right] + \gamma\, y(n-1) \qquad (5\text{-}12)$$

where $x(n)$ is the current input sample; $x(n\text{-}1)$ is the previous input; $y(n\text{-}1)$ is the previous output; and $y(n)$ is the current output. The filter coefficients α and γ are defined in Equation (5-8), as well as in (5-13b) where an alternate form of γ is given. Note that the coefficient β has been eliminated in Equation (5-13b).

$$y(n) = \alpha\left[x(n) - x(n-1)\right] + \gamma\, y(n-1) \qquad (5\text{-}13a)$$

$$\gamma = \frac{\cos\theta_c}{1 + \sin\theta_c} \qquad \alpha = (1+\gamma)/2 \qquad (5\text{-}13b)$$

where, $\theta \equiv 2\pi f / f_s \qquad \theta_c \equiv 2\pi f_c / f_s$

Difference equation of the first-order IIR high-pass filter with coefficient formulas

First-Order High-Pass Filter

The difference equation above can be implemented in C, as shown below. The delay elements, represented as rectangles in *Figure 5-3*, are implemented in the following example code by setting the program variable x1 = xn and y1 = yn at the end of the routine. In this way, x1 corresponds to *x(n-1)* and y1 corresponds to *y(n-1)*:

```cpp
//
// cook_highpass1.cpp
//
// Implementation of a simple 1st Order Highpass Filter Stage,
// class Chighpass1FilterStage

#include "cook.h"

Chighpass1FilterStage::Chighpass1FilterStage()
{
        x1 = 0;
        y1 = 0;
}

Chighpass1FilterStage::~Chighpass1FilterStage()
{
}

void Chighpass1FilterStage::execute_filter_stage()
{
        y = alpha * (x - x1) + gamma * y1 ;
        x1 = x;
        y1 = y;
}
```

6

Second-Order Low-Pass Filter

Analog Filter Network

A low-pass filter can be implemented using the RCL network shown in *Figure 6-1*. Note that other combinations of R, C, and L can also lead to low-pass filter characteristics, but will have different expressions for the damping factor d.

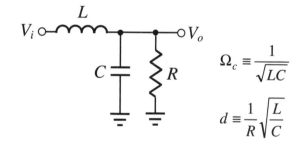

Figure 6-1. RCL low-pass filter network

$$\Omega_c \equiv \frac{1}{\sqrt{LC}}$$

$$d \equiv \frac{1}{R}\sqrt{\frac{L}{C}}$$

DSP Filters

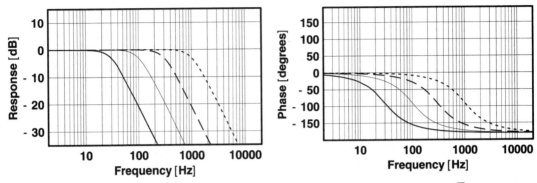

Figure 6-2. Gain response and phase plots for analog low-pass filter with $d = \sqrt{2}$: solid line, $f_c=30$; thin solid line, $f_c=100$; dashed line, $f_c=300$; and dotted line, $f_c=1000$

The R, C, and L complex impedances are dependent on the Laplace variable s as:

$$Z_R = R \qquad Z_C = \frac{1}{sC} \qquad Z_L = sL \qquad (6\text{-}1)$$

where a consistent set of units of measure include *ohms, farads, henrys,* and *Hertz*.

The ratio of the output V_o to input V_i in the circuit of *Figure 6-1* is equivalent to the ratio of the parallel resistor and capacitor leg to the total resistance of the circuit:

$$\frac{V_o}{V_i} = \frac{Z_R \| Z_C}{Z_L + Z_R \| Z_C} \qquad (6\text{-}2a)$$

Combining the expressions for complex impedance from Equations (6-1) for the R, C, and L components, results in the filter *transfer function* H(s):

$$H(s) \equiv \frac{V_o}{V_i} = \frac{R}{RCLs^2 + Ls + R} \qquad (6\text{-}2b)$$

where H(s) is defined as the ratio of V_o to input V_i. H(s) can be algebraically manipulated by dividing both the numerator and denominator by R and by LC:

Second-Order Low-Pass Filter

$$H(s) = \frac{\frac{1}{LC}}{s^2 + \frac{\sqrt{L/R^2C}}{\sqrt{LC}}s + \frac{1}{LC}} \quad (6\text{-}2c)$$

The result is a simplified representation of the complex low-pass filter transfer function:

$$H(s) = \frac{\Omega_c^2}{s^2 + d\,\Omega_c s + \Omega_c^2} \quad (6\text{-}3)$$

where,

$$\Omega_c \equiv \frac{1}{\sqrt{LC}} \qquad d \equiv \frac{1}{R}\sqrt{\frac{L}{C}} \quad (6\text{-}4)$$

An expression for the second-order low-pass filter response, which may be more useful than Equation (6-3), is the set of formulas describing the frequency magnitude response $G(f)$ and the phase response $\phi(f)$:

$$G(f) = \frac{f_c^2}{\sqrt{(f_c^2 - f^2)^2 + d^2 f_c^2 f^2}} \qquad \phi(f) = 180\left[\frac{1}{\pi}\tan^{-1}\left(\frac{f_c^2 - f^2}{d\,f_c\,f}\right) - \frac{1}{2}\right] \quad (6\text{-}5a)$$

$$f_c \equiv \frac{1}{2\pi\sqrt{LC}} \qquad d \equiv \frac{1}{R}\sqrt{\frac{L}{C}} \quad (6\text{-}5b)$$

Magnitude and phase response of the analog second-order low-pass filter

Figure 6-2 shows several examples of gain and phase response curves for the analog second-order low-pass filter using several values of cutoff frequency with $d = \sqrt{2}$: thick solid line, f_c = 30 Hz; thin solid line, f_c = 100 Hz; dashed line, f_c = 300 Hz; and dotted line, f_c = 1000 Hz. Note that the gain response curve of Figure 6-2 is *maximally flat* before it begins to roll off above the cutoff frequency when $d = \sqrt{2}$. The magnitude of the gain and phase at the cutoff frequency can easily be determined by substituting $f = f_c$ in Equations (6-5a):

DSP Filters

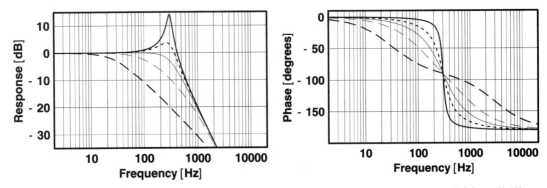

Figure 6-3. Gain response and phase plots for analog low-pass filter with fc = 300: solid line, d = 0.2; dotted line, $d = \sqrt{2}$; thin solid line, $d = \sqrt{2}$; thin dashed line, d = 3; dashed line, d = 10

$$G(f_c) = \frac{f_c^2}{\sqrt{(f_c^2 - f_c^2)^2 + d^2 f_c^4}} = \boxed{\frac{1}{d}} \quad (6\text{-}6a)$$

$$\phi(f_c) = 180\left[\frac{1}{\pi}\tan^{-1}\left(\frac{f_c^2 - f_c^2}{d\, f_c\, f_c}\right) - \frac{1}{2}\right] = \boxed{-90°} \quad (6\text{-}6b)$$

Gain and phase response of analog low-pass filter at its cutoff frequency

Figure 6-3 shows the gain $G(f)$ and phase $\phi(f)$ plots for the second-order low-pass RCL network of *Figure 6-1* for various values of damping factor d with $f_c = 300$: solid line, $d = 0.2$; dotted line, $d = 1/\sqrt{2}$; thin solid line, $d = \sqrt{2}$; thin dashed line, $d = 3$; and dashed line, $d = 10$. The damping factor d determines the precise amplitude and position of the peak of the gain response curve. The frequency f_m corresponding to the peak amplitude and the magnitude of this peak, can be found by taking the derivative of $G(f)$ with respect to f from Equation (6-5) and setting it equal to 0:

Second-Order Low-Pass Filter

$$\frac{d}{df}G(f) = \frac{d}{df}\frac{f_c^2}{\sqrt{(f_c^2 - f^2)^2 + d^2 f_c^2 f^2}} = 0 \Rightarrow f_m = \sqrt{\frac{2-d^2}{2}} f_c \quad (6\text{-}7)$$

Substituting f_m from above into Equations (6-5) determines the precise amplitude and position of the peak of the gain curve:

$$f_m = \sqrt{\frac{2-d^2}{2}} f_c \quad (6\text{-}8a)$$

$$G(f_m) = \frac{f_c^2}{\sqrt{(f_c^2 - f_m^2)^2 + d^2 f_c^2 f_m^2}} = \frac{2}{d\sqrt{4-d^2}} \quad (6\text{-}8b)$$

$$\phi(f_m) = 180\left[\frac{1}{\pi}\tan^{-1}\left(\frac{f_c^2 - f_m^2}{d\, f_c\, f_m}\right) - \frac{1}{2}\right] = 180\left[\frac{1}{\pi}\tan^{-1}\left(\frac{d}{\sqrt{4-2d^2}}\right) - \frac{1}{2}\right] \quad (6\text{-}8c)$$

Gain and phase values at the peak gain frequency for the analog low-pass filter

Digital Filter Network

An IIR second-order digital low-pass filter is constructed by implementing a *running average* on a stream of sampled data, using two previous inputs and two previous outputs for each sample time.

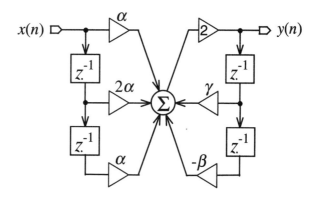

Figure 6-4. Digital low-pass filter network

DSP Filters

The transfer function of the low-pass network in *Figure 6-4* describes its frequency response characteristics, similar to the s-domain transfer function of Equation (6-3), but is now a function of the z-transform:

$$H(z) = \frac{\alpha\left(1 + 2z^{-1} + z^{-2}\right)}{\tfrac{1}{2} - \gamma z^{-1} + \beta z^{-2}} \qquad (6\text{-}9)$$

The z variable of the z-transform represents a point in the complex plane where the digital filter response is found along the unit circle at $z = e^{j 2\pi f / f_s}$, where f_s is the sample frequency.

The coefficients α, β, and γ in Equation (6-9) are directly related to the filter damping factor d and cutoff frequency f_c, by:

$$\beta = \frac{1}{2}\,\frac{1 - \tfrac{d}{2}\sin(2\pi f_c / f_s)}{1 + \tfrac{d}{2}\sin(2\pi f_c / f_s)}$$

$$\gamma = \left(\tfrac{1}{2} + \beta\right)\cos(2\pi f_c / f_s) \qquad (6\text{-}10)$$

$$\alpha = \left(\tfrac{1}{2} + \beta - \gamma\right)/4$$

The magnitude of the complex z-transform is found by taking the absolute value of H(z): $G(f) = |H(z)| = \sqrt{H(z) H^*(z)}$. The phase response $\phi(f)$ is the phase angle between the imaginary and real components of H(z) and is computed by taking the inverse tangent of the ratio of those components: $\phi(f) = \tan^{-1}[\mathrm{Im}\{H(z)\}/\mathrm{Re}\{H(z)\}]$. Substituting $z = e^{j 2\pi f / f_s}$ into these relationships and applying trigonometric identities with a substantial effort at algebraic reduction results in the formulas of Equations (6-11).

Figure 6-5 shows several examples of gain and phase response curves for the digital second-order low-pass filter, using several values of cutoff frequency, with $d = \sqrt{2}$: thick solid line, f_c = 30 Hz; thin solid line, f_c = 100 Hz; dashed line, f_c = 300 Hz; and dotted line, f_c = 1000 Hz. Note that the gain response curves of *Figure 6-5* are *maximally flat* with $d = \sqrt{2}$, as in the case of the analog response curves of *Figure 6-2*. The differences between the response of the analog filter of *Figure 6-2* and the digital

Second-Order Low-Pass Filter

$$G(f) = \frac{(1+\cos\theta)(1-\cos\theta_c)}{\sqrt{(d\sin\theta\sin\theta_c)^2 + 4(\cos\theta-\cos\theta_c)^2}} \quad (6\text{-}11a)$$

$$\phi(f) = 180\left[\frac{1}{\pi}\tan^{-1}\left(\frac{2(\cos\theta-\cos\theta_c)}{d\sin\theta\sin\theta_c}\right) - \frac{1}{2}\right] \quad (6\text{-}11b)$$

where, $\theta \equiv 2\pi f / f_s \quad \theta_c \equiv 2\pi f_c / f_s$

Magnitude and phase response of the digital second-order low-pass filter

filter of *Figure 6-5* are that the digital filter response goes to zero (minus infinity on a dB plot) at the Nyquist frequency $f_n = f_s/2$, whereas in this example, f_s = 11025 Hz and f_n = 5512.5 Hz. If the sample frequency is set to f_s = 44100 Hz or higher, the digital frequency response plots corresponding to the parameter values used in *Figure 6-5* will appear almost identical to those of *Figure 6-2*.

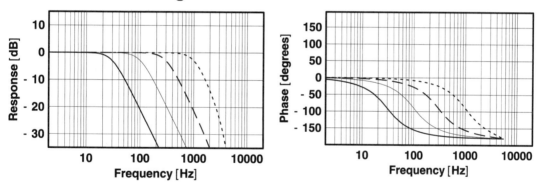

Figure 6-5. Gain and phase of digital low-pass filter with $d = \sqrt{2}$ and f_s = 11025: solid line, f_c = 30; thin solid line, f_c = 100; dashed line, f_c = 300; dotted line, f_c = 1000

As the sample frequency f_s of an IIR filter is increased, the frequency response approaches that of its analog filter counterpart. Likewise, as the cutoff frequency f_c and operating frequency f is decreased, the digital frequency response approaches that of its analog filter counterpart.

DSP Filters

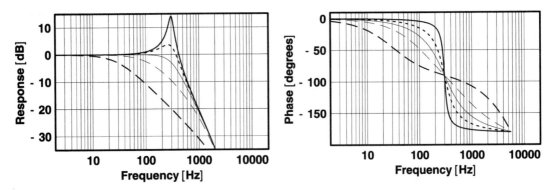

Figure 6-6. Gain phase for digital low-pass filter with $f_c = 300$ and $f_s = 11025$: solid line, $d = 0.2$; dotted line, $d = 1/\sqrt{2}$; thin solid line, $d = \sqrt{2}$; thin dashed line, $d = 3$; dashed line, $d = 10$

This fundamental principle can again be demonstrated by recalling that the trigonometric functions, cos x and sin x, can be represented by:

$$\left. \begin{array}{c} \cos\theta \to 1 - \tfrac{1}{2}\theta^2 \\ \sin\theta \to \theta \end{array} \right\} \text{ for } \theta \to 0 \qquad (6\text{-}12)$$

and then substituting in Equation (6-11a):

$$G(f) \to \frac{(1 + 1 - \tfrac{1}{2}\theta^2)(1 - 1 + \tfrac{1}{2}\theta_c^2)}{\sqrt{(d\theta\theta_c)^2 + 4(1 - \tfrac{1}{2}\theta^2 - 1 + \tfrac{1}{2}\theta_c^2)^2}} \quad \text{ for } \theta \to 0$$

$$= \frac{(2 - \tfrac{1}{2}\theta^2)(\tfrac{1}{2}\theta_c^2)}{\sqrt{(d\theta\theta_c)^2 + (-\theta^2 + \theta_c^2)^2}} \qquad (6\text{-}13)$$

$$\approx \frac{\theta_c^2}{\sqrt{(d\theta\theta_c)^2 + (\theta_c^2 - \theta^2)^2}} = \boxed{\frac{f_c^2}{\sqrt{(d f f_c)^2 + (f_c^2 - f^2)^2}}}$$

Comparing the result of Equation (6-13) with the analog gain response of Equation (6-5a), it can be seen that they are equivalent. A similar result is obtained by substituting Equation (6-12) in the IIR phase formula of Equation (6-11b) — it will be identical to the analog low-pass phase formula of Equation (6-5b).

Second-Order Low-Pass Filter

Difference Equation

In order to implement the digital IIR filter, a *difference equation* is used:

$$y(n) = 2\{\alpha\,[x(n) + 2x(n-1) + x(n-2)] + \gamma\,y(n-1) - \beta\,y(n-2)\} \quad (6\text{-}14)$$

where $x(n)$ is the current input sample; $x(n\text{-}1)$ is the previous input; $x(n\text{-}2)$ is the previous previous input; $y(n\text{-}1)$ is the previous output; $y(n\text{-}2)$ is the previous previous output; and $y(n)$ is the current output. The filter coefficients α, β, and γ are defined in Equation (6-10), as well as in (6-15b).

$$y(n) = 2\{\alpha\,[x(n) + 2x(n-1) + x(n-2)] + \gamma\,y(n-1) - \beta\,y(n-2)\} \quad (6\text{-}15a)$$

$$\beta = \frac{1}{2}\frac{1 - \frac{d}{2}\sin\theta_c}{1 + \frac{d}{2}\sin\theta_c} \qquad \gamma = \left(\tfrac{1}{2} + \beta\right)\cos\theta_c \qquad \alpha = \left(\tfrac{1}{2} + \beta - \gamma\right)/4 \quad (6\text{-}15b)$$

where, $\theta \equiv 2\pi f / f_s \qquad \theta_c \equiv 2\pi f_c / f_s$

Difference equation of the digital second-order low-pass filter with coefficient formulas

As discussed in previous chapters, there are several ways in which to interpret the second order low-pass difference equation. One of the fundamental interpretations is that Equation (6-14) is the result of applying the inverse z-transform to $H(z)$ as discussed in Chapter 3, Equations (3-8) through (3-11).

DSP Filters

The following C code implements the low-pass difference equation:

```cpp
//
// cook_lowpass.cpp
//
// Implementation of a simple Low-pass filter Stage,
// class CLowpassFilterStage

#include "cook.h"

CLowpassFilterStage::CLowpassFilterStage()
{
        x1 = 0;
        x2 = 0;
        y1 = 0;
        y2 = 0;
}

CLowpassFilterStage::~CLowpassFilterStage()
{
}

void CLowpassFilterStage::execute_filter_stage()
{
        y = 2 * ((alpha * (x + 2.0 * x1+ x2) + gamma * y1 - beta * y2);
        x2 = x1;
        x1 = x;
        y2 = y1;
        y1 = y;
}
```

7
Second-Order High-Pass Filter

Analog Filter Network

An analog high-pass filter can be implemented using the RCL network described in *Figure 7-1*. As with the low-pass filter of the previous chapter, other combinations of *R*, *C*, and *L* can also lead to second-order high-pass filter characteristics, but will have a different dependence of damping factor *d* on the circuit components.

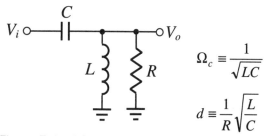

Figure 7-1. RCL high-pass filter network

$$\Omega_c \equiv \frac{1}{\sqrt{LC}}$$

$$d \equiv \frac{1}{R}\sqrt{\frac{L}{C}}$$

The filter of *Figure 7-1* is expressed as the ratio of the output V_o to input V_i and is equivalent to the ratio of the parallel resistor and inductor leg to the total resistance of the network:

DSP Filters

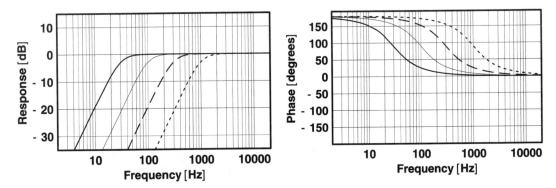

Figure 7-2. Gain response and phase plots for analog high-pass filter with $d = \sqrt{2}$: solid line, $f_c = 30$; thin solid line, $f_c = 100$; dashed line, $f_c = 300$; dotted line, $f_c = 1000$

$$\frac{V_o}{V_i} = \frac{Z_R \| Z_L}{Z_C + Z_R \| Z_L} \quad (7\text{-}1a)$$

where,

$$Z_R = R \qquad Z_C = \frac{1}{sC} \qquad Z_L = sL \quad (7\text{-}1b)$$

Combining the expressions for complex impedance from Equations (7-1b) with the voltage divider formula of Equation (7-1a), results in a high-pass filter transfer function:

$$H(s) \equiv \frac{V_o}{V_i} = \frac{sLR}{\frac{R+sL}{sC} + sLR} \quad (7\text{-}2a)$$

$H(s)$ can be algebraically manipulated by dividing both the numerator and denominator by LR and multiplied by s :

$$H(s) = \frac{s^2}{s^2 + \frac{\sqrt{L/R^2C}}{\sqrt{LC}} s + \frac{1}{LC}} \quad (7\text{-}2b)$$

The result is a simplified representation of the complex high-pass filter transfer function:

$$H(s) = \frac{s^2}{s^2 + d\,\Omega_c s + \Omega_c^2} \quad (7\text{-}3)$$

where,

$$\Omega_c \equiv \frac{1}{\sqrt{LC}} \qquad d \equiv \frac{1}{R}\sqrt{\frac{L}{C}} \qquad (7\text{-}4)$$

The magnitude response $G(f)$ of the complex transfer function is found by taking the absolute value of $H(s)$: $G(f) = |H(s)| = \sqrt{H(s)H^*(s)}$. The phase response $\phi(f)$ is the angle between the imaginary and real components of $H(s)$, found by taking the inverse tangent of that ratio: $\phi(f) = \tan^{-1}[\text{Im}\{H(s)\}/\text{Re}\{H(s)\}]$. Substituting $s = j2\pi f$ into these relationships and applying trigonometric identities with a generous amount of algebraic simplification results in the formulas of Equations (7-5):

$$G(f) = \frac{f^2}{\sqrt{(f_c^2 - f^2)^2 + d^2 f_c^2 f^2}} \qquad \phi(f) = 180\left[\frac{1}{\pi}\tan^{-1}\left(\frac{f_c^2 - f^2}{d f_c f}\right) + \frac{1}{2}\right] \qquad (7\text{-}5a)$$

$$f_c \equiv \frac{1}{2\pi\sqrt{LC}} \qquad d \equiv \frac{1}{R}\sqrt{\frac{L}{C}} \qquad (7\text{-}5b)$$

Magnitude and phase response of the analog second-order high-pass filter

Figure 7-2 shows several examples of gain and phase response curves for the analog second-order high-pass filter using several values of cutoff frequency, with $d = \sqrt{2}$: thick solid line, f_c = 30 Hz; thin solid line, f_c = 100 Hz; dashed line, f_c = 300 Hz; and dotted line, f_c = 1000 Hz. As in the low-pass case of the previous chapter, the gain response of Figure 7-2 is *maximally flat* before it begins to roll off below the cutoff frequency when $d = \sqrt{2}$. The magnitude of the gain and phase at the cutoff frequency can easily be determined by substituting $f = f_c$ in Equations (7-5a):

DSP Filters

$$G(f_c) = \frac{f_c^2}{\sqrt{(f_c^2 - f_c^2)^2 + d^2 f_c^4}} = \boxed{\frac{1}{d}} \quad (7\text{-}6a)$$

$$\phi(f_c) = 180\left[\frac{1}{\pi}\tan^{-1}\left(\frac{f_c^2 - f_c^2}{d\, f_c\, f_c}\right) + \frac{1}{2}\right] = \boxed{90°} \quad (7\text{-}6b)$$

Gain and phase response of the analog high-pass filter at its cutoff frequency

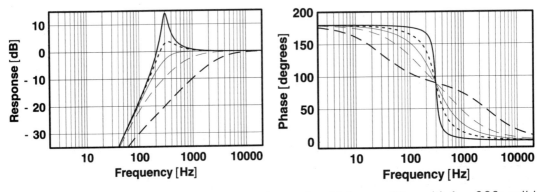

Figure 7-3. Gain response and phase plots for analog high-pass filter with $f_c = 300$: solid line, $d = 0.2$; dotted line, $d = 1/\sqrt{2}$; thin solid line, $d = \sqrt{2}$; thin dashed line, $d = 3$; dashed line, $d = 10$

Figure 7-3 shows the gain $G(f)$ and phase $\phi(f)$ plots for the second-order high-pass RCL network of Figure 7-1 for various values of damping factor d with $f_c = 300$: solid line, $d = 0.2$; dotted line, $d = 1/\sqrt{2}$; thin solid line, $d = \sqrt{2}$; thin dashed line, $d = 3$; and dashed line, $d = 10$. The damping factor d determines the precise amplitude and position of the peak of the gain response curve. The frequency f_m corresponding to the peak amplitude and the magnitude of this peak can be found by taking the derivative of $G(f)$ with respect to f, from Equation (7-5) and setting it equal to 0:

Second-Order High-Pass Filter

$$\frac{d}{df}G(f) = \frac{d}{df}\frac{f^2}{\sqrt{(f_c^2 - f^2)^2 + d^2 f_c^2 f^2}} = 0 \Rightarrow f_m = \sqrt{\frac{2}{2-d^2}} f_c \quad (7\text{-}7)$$

Substituting f_m from above into Equation (7-5) determines the precise magnitude and phase values at the position of the peak of the magnitude (gain) curve:

$$f_m = \sqrt{\frac{2}{2-d^2}} f_c \quad (7\text{-}8a)$$

$$G(f_m) = \frac{f_m^2}{\sqrt{(f_c^2 - f_m^2)^2 + d^2 f_c^2 f_m^2}} = \frac{2}{d\sqrt{4-d^2}} \quad (7\text{-}8b)$$

$$\phi(f_m) = 180\left[\frac{1}{\pi}\tan^{-1}\left(\frac{f_c^2 - f_m^2}{d f_c f_m}\right) + \frac{1}{2}\right] = 180\left[\frac{1}{\pi}\tan^{-1}\left(\frac{d}{\sqrt{4-2d^2}}\right) + \frac{1}{2}\right] \quad (7\text{-}8c)$$

Gain and phase values at the peak gain frequency for the analog high-pass filter

Digital Filter Network

Like the s-domain transfer function of Equation (7-3), the z-transform function $H(z)$ of the high-pass network shown in Figure 7-4 describes its filter response characteristics:

$$H(z) = \frac{\alpha\left(1 - 2z^{-1} + z^{-2}\right)}{\frac{1}{2} - \gamma z^{-1} + \beta z^{-2}} \quad (7\text{-}9)$$

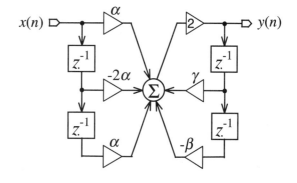

Figure 7-4. Digital high-pass filter network

DSP Filters

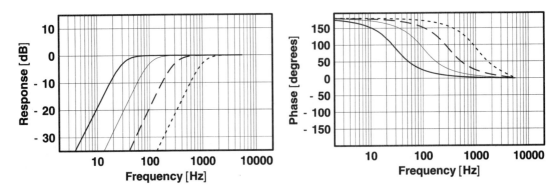

Figure 7-5. Gain and phase of digital high-pass filter with $d = \sqrt{2}$ and fs = 11025: solid line, f_c = 30; thin solid line, f_c = 100; dashed line, fc = 300; dotted line, f_c = 1000

The coefficients α, β, and γ in Equation (7-9) are directly related to the filter damping factor d and cutoff frequency f_c, by:

$$\beta = \frac{1}{2}\frac{1 - \frac{d}{2}\sin(2\pi f_c / f_s)}{1 + \frac{d}{2}\sin(2\pi f_c / f_s)}$$

$$\gamma = \left(\tfrac{1}{2} + \beta\right)\cos(2\pi f_c / f_s) \quad (7\text{-}10)$$

$$\alpha = \left(\tfrac{1}{2} + \beta + \gamma\right)/4$$

Figure 7-5 shows several examples of gain and phase response curves for the digital second-order high-pass filter using several values of cutoff frequency, with $d = \sqrt{2}$: thick solid line, f_c = 30 Hz; thin solid line, f_c = 100 Hz; dashed line, f_c = 300 Hz; and dotted line, f_c = 1000 Hz. Note that the gain response curves of Figure 7-5 are *maximally flat* with $d = \sqrt{2}$, as in the case of the analog response curves of Figure 7-2.

Note the similarities between the response of the analog filter of Figure 7-2 and the digital filter of Figure 7-5. Both the magnitude and phase response follow the same curves below the Nyquist frequency $f_n = f_s/2$. Above the Nyquist, the digital filter response is a mirror image of the response below f_n, but is generally not shown on a filter response

Second-Order High-Pass Filter

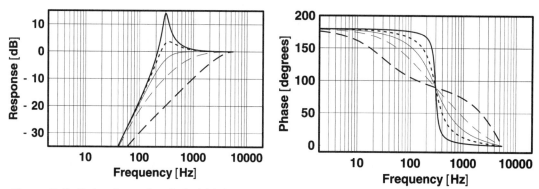

Figure 7-6. Gain phase for digital high-pass filter with fc = 300 and fs = 11025: solid line, d = 0.2; dotted line, $d = 1/\sqrt{2}$; thin solid line, $d = \sqrt{2}$; thin dashed line, d = 3; dashed line, d = 10

plot. If the output of the digital filter is passed through a near ideal anti-aliasing filter, such as those found internal to most audio DACs (see Chapter 3), the magnitude response above the Nyquist will be very near zero (or large negative values on a dB plot).

The magnitude G(f) is calculated by finding the absolute value of H(z): $G(f) = |H(z)| = \sqrt{H(z)H^*(z)}$. The phase response φ(f) is the angle between the imaginary and real components of H(z), found by taking the inverse tangent of that ratio: $\phi(f) = \tan^{-1}[\text{Im}\{H(z)\}/\text{Re}\{H(z)\}]$. Substituting $z = e^{j2\pi f/f_s}$ into these relationships and, as in the previous chapter, applying trigonometric identities with a sufficient endeavor at algebraic reduction, results in the formulas of Equations (7-11):

$$G(f) = \frac{(1-\cos\theta)(1+\cos\theta_c)}{\sqrt{(d\sin\theta\sin\theta_c)^2 + 4(\cos\theta - \cos\theta_c)^2}} \quad (7\text{-}11\text{a})$$

$$\phi(f) = 180\left[\frac{1}{\pi}\tan^{-1}\left(\frac{2(\cos\theta - \cos\theta_c)}{d\sin\theta\sin\theta_c}\right) + \frac{1}{2}\right] \quad (7\text{-}11\text{b})$$

where, $\theta \equiv 2\pi f / f_s$ $\quad \theta_c \equiv 2\pi f_c / f_s$

Magnitude and phase response of the digital second-order high-pass filter

DSP Filters

This similarity between analog and digital response functions can again be demonstrated as it was in the second-order low-pass case, by recalling that the trigonometric functions, cos x and sin x, obey the following limits:

$$\left.\begin{array}{l}\cos\theta \to 1-\tfrac{1}{2}\theta^2 \\ \sin\theta \to \theta\end{array}\right\} \text{ for } \theta \to 0 \qquad (7\text{-}12)$$

Substituting Equation (7-12) in Equation (7-11a) results in:

$$G(f) \to \frac{\left(1-1+\tfrac{1}{2}\theta^2\right)\left(1+1-\tfrac{1}{2}\theta_c^2\right)}{\sqrt{(d\theta\theta_c)^2 + 4\left(1-\tfrac{1}{2}\theta^2 - 1 + \tfrac{1}{2}\theta_c^2\right)^2}} \quad \text{for } \theta \to 0$$

$$= \frac{\left(\tfrac{1}{2}\theta^2\right)(2)}{\sqrt{(d\theta\,\theta_c)^2 + \left(-\theta^2 + \theta_c^2\right)^2}} \qquad (7\text{-}13)$$

$$\approx \frac{\theta^2}{\sqrt{(d\theta\,\theta_c)^2 + \left(\theta_c^2 - \theta^2\right)^2}} = \boxed{\frac{f^2}{\sqrt{(d\,f\,f_c)^2 + \left(f_c^2 - f^2\right)^2}}}$$

Comparing the result of Equation (7-13) with the analog gain response of Equation (7-5a), it can be seen that they are equivalent. A similar result is obtained by substituting Equation (7-12) in the IIR phase formula of Equation (7-11b) — it will be identical to the analog high-pass phase formula of Equation (7-5b) as $\theta \to 0$.

Difference Equation

In order to implement the digital IIR filter, a *difference equation* is used. As previously mentioned, the IIR filter implementation is simply a running average of the input and output data:

$$y(n) = 2\{\alpha\,[x(n) - 2x(n-1) + x(n-2)] + \gamma\,y(n-1) - \beta\,y(n-2)\} \qquad (7\text{-}14)$$

Second-Order High-Pass Filter

where *x(n)* is the current input sample; *x(n-1)* is the previous input; *x(n-2)* is the previous previous input; *y(n-1)* is the previous output; *y(n-2)* is the previous previous output; and *y(n)* is the current output. The filter coefficients α, β, and γ are defined in Equation (7-10), as well as in (7-15b).

$$y(n) = 2\{\alpha [x(n) - 2x(n-1) + x(n-2)] + \gamma\, y(n-1) - \beta\, y(n-2)\} \quad (7\text{-}15a)$$

$$\beta = \frac{1}{2}\left(\frac{1 - \frac{1}{2}d\sin\theta_c}{1 + \frac{1}{2}d\sin\theta_c}\right) \qquad \gamma = \left(\tfrac{1}{2} + \beta\right)\cos\theta_c \qquad \alpha = \left(\tfrac{1}{2} + \beta + \gamma\right)/4 \quad (7\text{-}15b)$$

$$\text{where,} \quad \theta \equiv 2\pi f / f_s \qquad \theta_c \equiv 2\pi f_c / f_s$$

Difference equation of the digital second-order high-pass filter with coefficient formulas

Example C code to implement Equation (7-14) follows.

```
//
// cook_highpass.cpp
//
// Implementation of a simple Highpass Filter Stage,
// class CHighpassFilterStage

#include "cook.h"

CHighpassFilterStage::CHighpassFilterStage()
{
        x1 = 0;
        x2 = 0;
        y1 = 0;
        y2 = 0;
}
```

DSP Filters

```
CHighpassFilterStage::~CHighpassFilterStage()
{
}

void CHighpassFilterStage::execute_filter_stage()
{
    y = 2 * ((alpha * (x - 2.0 * x1 + x2) + gamma * y1 - beta * y2);
    x2 = x1;
    x1 = x;
    y2 = y1;
    y1 = y;
}
```

8
Second-Order Bandpass Filter

Analog Filter Network

A bandpass filter can be implemented using the RCL network shown in *Figure 8-1*, where the impedances associated with perfect R, C, and L components are:

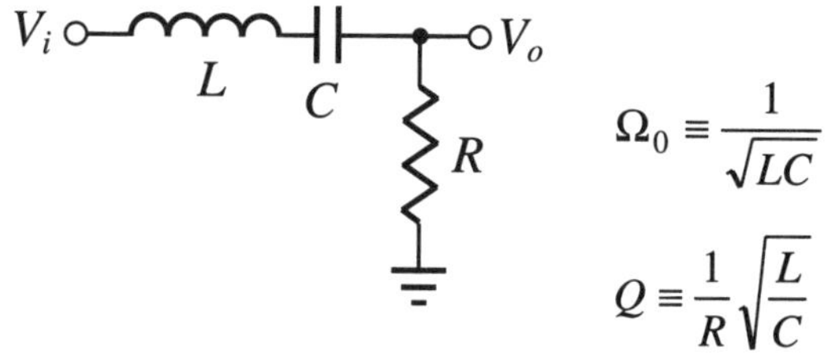

Figure 8-1. RCL bandpass filter network

$$\Omega_0 \equiv \frac{1}{\sqrt{LC}}$$

$$Q \equiv \frac{1}{R}\sqrt{\frac{L}{C}}$$

$$Z_R = R \quad Z_C = \frac{1}{sC} \quad Z_L = sL \quad (8\text{-}1)$$

The units of measure (*mks* system) of these components are: resistance R, in ohms; capacitance C, in farads; inductance L, in henrys; and frequency f, in Hertz.

DSP Filters

The filter of *Figure 8-1* is expressed as the ratio of the output V_o to input V_i and is equivalent to the ratio of the series resistor to the total resistance of the network:

$$\frac{V_o}{V_i} = \frac{Z_R}{Z_L + Z_C + Z_R} \quad (8\text{-}2a)$$

The expressions for complex impedance from Equations (8-1) for the R, C, and L components can be combined with Equation (8-2a), resulting in a complex analog filter *transfer function H(s)*:

$$H(s) \equiv \frac{V_o}{V_i} = \frac{R}{Ls + 1/sC + R} \quad (8\text{-}2b)$$

where H(s) is defined as the ratio of V_o to input V_i. Dividing both the numerator and denominator by L and multiplying by s, results in:

$$H(s) = \frac{sR/L}{s^2 + sR/L + 1/LC}$$

$$= \frac{\dfrac{sR}{\sqrt{LC}}\sqrt{\dfrac{C}{L}}}{s^2 + \dfrac{sR}{\sqrt{LC}}\sqrt{\dfrac{C}{L}} + \dfrac{1}{LC}} \quad (8\text{-}2c)$$

and finally,

$$H(s) = \frac{s\Omega_0/Q}{s^2 + s\Omega_0/Q + \Omega_0^2} \quad (8\text{-}3)$$

with,

$$\Omega_0 \equiv \frac{1}{\sqrt{LC}} \quad Q \equiv \frac{1}{R}\sqrt{\frac{L}{C}} \quad (8\text{-}4)$$

where the damping factor $d = 1/Q$.

The magnitude G(f) and the phase angle φ(f) of H(s) are summarized in Equations (8-5).

Second-Order Bandpass Filter

$$G(f) = \frac{f_0 f / Q}{\sqrt{(f_0^2 - f^2)^2 + (f_0 f / Q)^2}} \qquad \phi(f) = 180\left[\frac{1}{\pi} \tan^{-1}\left(\frac{f_0^2 - f^2}{f_0 f / Q}\right)\right] \qquad (8\text{-}5a)$$

$$f_0 \equiv \frac{1}{2\pi\sqrt{LC}} \qquad Q \equiv \frac{1}{R}\sqrt{\frac{L}{C}} \qquad (8\text{-}5b)$$

Magnitude and phase response of the analog second-order bandpass filter

Figure 8-2 shows several examples of gain and phase response curves for the analog second-order bandpass filter, using several values of center frequency, with $Q = 1$: thick solid line, $f_0 = 30$ Hz; thin solid line, $f_0 = 100$ Hz; dashed line, $f_0 = 300$ Hz; and dotted line, $f_0 = 1000$ Hz. The roll-off on either side of f_0 goes as -6 dB/octave or -20 dB/decade.

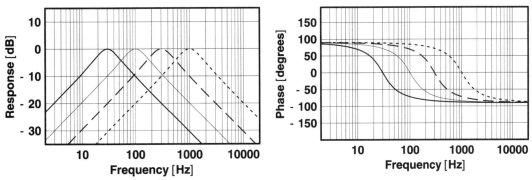

Figure 8-2. Gain response and phase plots for analog bandpass filter with Q = 1: solid line, $f_0 = 30$; thin solid line, $f_0 = 100$; dashed line, $f_0 = 300$; dotted line, $f_0 = 1000$

The magnitude of the gain and phase at the center frequency can easily be determined by substituting $f = f_0$ in Equations (8-5a):

Gain and phase response of analog bandpass filter at its center frequency

$$G(f_0) = \frac{f_0 f_0 / Q}{\sqrt{(f_0^2 - f_0^2)^2 + (f_0 f_0 / Q)^2}} = 1 \qquad (8\text{-}6a)$$

$$\phi(f_0) = 180\left[\frac{1}{\pi} \tan^{-1}\left(\frac{f_0^2 - f_0^2}{f_0 f_0 / Q}\right)\right] = 0° \qquad (8\text{-}6b)$$

81

DSP Filters

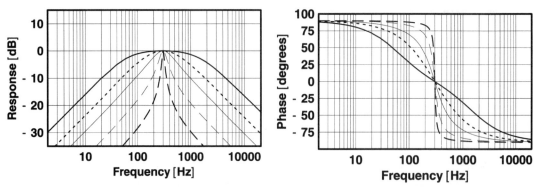

Figure 8-3. Gain response and phase plots for analog bandpass filter with $f_0 = 300$: solid line, $Q = 0.2$; dotted line, $Q = 0.5$; thin solid line, $Q = 1$; thin dashed line, $Q = 3$; dashed line, $Q = 10$

Figure 8-3 shows the gain $G(f)$ and phase $\phi(f)$ plots for the second-order bandpass RCL network of *Figure 8-1* for various values of *quality factor* Q with $f_0 = 300$: solid line, $Q = 0.2$; dotted line, $Q = 0.5$; thin solid line, $Q = 1$; thin dashed line, $Q = 3$; and dashed line, $Q = 10$.

The quality factor Q is inversely proportional to the frequency bandwidth Δf of the gain response curve:

$$Q \equiv \frac{f_0}{\Delta f} \quad (8\text{-}7a)$$

where $\Delta f = f_2 - f_1$. The center frequency f_0 is equal to the geometric mean of f_1 and f_2:

$$f_0 = \sqrt{f_1 f_2} \quad (8\text{-}7b)$$

The *band-edge* frequencies f_1 and f_2 are defined by the following relationships shown in Equations (8-8a and 8-8b). The gain and phase values at the edge frequencies can be found by plugging in the expressions for f_1 and f_2 from Equations (8-8) into the gain and phase formulas from Equation (8-5), as shown in Equations 8-9:

Second-Order Bandpass Filter

$$f_1 = \frac{f_0}{2Q}\left(\sqrt{1+4Q^2} - 1\right) \tag{8-8a}$$

$$f_2 = \frac{f_0}{2Q}\left(\sqrt{1+4Q^2} + 1\right) \tag{8-8b}$$

where,

$$Q \equiv \frac{f_0}{\Delta f} \qquad \Delta f \equiv f_2 - f_1 \qquad f_0 = \sqrt{f_1 f_2}$$

Definition of quality factor Q and its relationship to bandwidth Δf

$$G(f_1) = \frac{f_0 f_1 / Q}{\sqrt{(f_0^2 - f_1^2)^2 + (f_0 f_1 / Q)^2}} = \frac{1}{\sqrt{2}} \tag{8-9a}$$

$$G(f_2) = \frac{f_0 f_2 / Q}{\sqrt{(f_0^2 - f_2^2)^2 + (f_0 f_2 / Q)^2}} = \frac{1}{\sqrt{2}} \tag{8-9b}$$

$$\phi(f_1) = 180\left[\frac{1}{\pi}\tan^{-1}\left(\frac{f_0^2 - f_1^2}{f_0 f_1 / Q}\right)\right] = 45° \tag{8-9c}$$

$$\phi(f_2) = 180\left[\frac{1}{\pi}\tan^{-1}\left(\frac{f_0^2 - f_2^2}{f_0 f_2 / Q}\right)\right] = -45° \tag{8-9d}$$

Gain and phase response of analog bandpass filter at the edge frequencies

Digital Filter Network

As detailed in Chapter 3, the z-transform of the discrete time domain network shown in Figure 8-4 gives the z-transfer function H(z). Like the s-transfer function H(s) of Equation (8-3), the z-transfer function of the bandpass network of Figure 8-4 describes its frequency response characteristics:

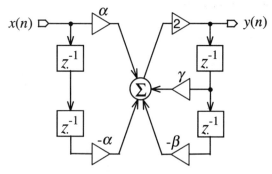

Figure 8-4. Digital bandpass filter network

$$H(z) = \frac{\alpha(1 - z^{-2})}{\frac{1}{2} - \gamma z^{-1} + \beta z^{-2}} \qquad (8\text{-}10)$$

Recall from previous discussion that the s variable of the Laplace transform (or s-transform) represents a point in the complex plane, such that the analog filter response characteristics are found along the vertical line determined by setting $s = j2\pi f$. In a similar manner, the z variable of the z-transform is again a point in the complex plane, but now the digital filter response is found along the unit circle at $z = e^{j2\pi f/f_s}$, where f_s is the sample frequency. The transformation from the s-domain to the z-domain is known as the *bilinear transformation*.

The coefficients α, β, and γ in Equation (8-10) are directly related to the filter quality factor Q and center frequency f_0, by:

$$\beta = \frac{1}{2}\left(\frac{1 - \tan\frac{\theta_0}{2Q}}{1 + \tan\frac{\theta_0}{2Q}}\right)$$

$$\gamma = \left(\tfrac{1}{2} + \beta\right)\cos(\theta_0)$$

$$\alpha = \left(\tfrac{1}{2} - \beta\right)/2$$

(8-11)

Second-Order Bandpass Filter

where $\theta_0 = 2\pi f_0 / f_s$, which is the *normalized center frequency*. The digital bandpass filter frequency response is the magnitude G(f) and the phase angle φ(f) of H(z):

$$G(f) = \frac{\sin\theta \, \tan\frac{\theta_0}{2Q}}{\sqrt{\left(\sin\theta \, \tan\frac{\theta_0}{2Q}\right)^2 + (\cos\theta - \cos\theta_0)^2}} \quad (8\text{-}12a)$$

$$\phi(f) = 180\left[\frac{1}{\pi}\tan^{-1}\left(\frac{\cos\theta - \cos\theta_0}{\sin\theta \, \tan\frac{\theta_0}{2Q}}\right)\right] \quad (8\text{-}12b)$$

where, $\theta \equiv 2\pi f / f_s$ $\quad \theta_0 \equiv 2\pi f_0 / f_s$

Magnitude and phase response of the digital second-order bandpass filter

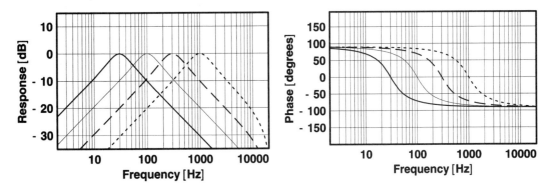

Figure 8-5. Gain and phase response of digital bandpass filter with Q = 1 and f_s = 44100: solid line, f_0 = 30; thin solid line, f_0 = 100; dash line, f_0 = 300; dotted line, f_0 = 1000

Figure 8-5 shows examples of gain and phase response curves for the digital second-order bandpass filter using several values of center frequency, with Q = 1: thick solid line, f_0 = 30 Hz; thin solid line, f_0 = 100

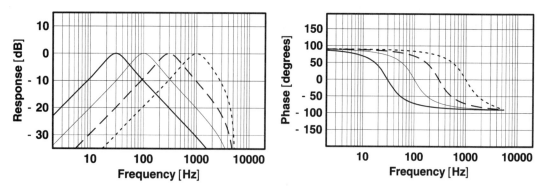

Figure 8-6. Gain and phase response of digital bandpass filter with Q = 1 and f_s = 11025: solid line, f_0 = 30; thin solid line, f_0 = 100; dash line, f_0 = 300; dotted line, f_0 = 1000

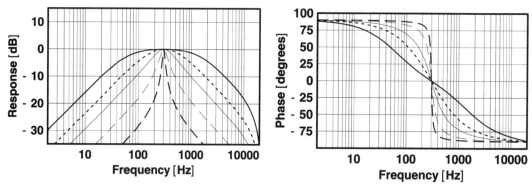

Figure 8-7. Gain response and phase plots for IIR bandpass filter with f_s = 44100 and f_0 = 300: solid line, Q = 0.2; dotted line, Q = 0.5; thin solid line, Q = 1; dashed line, Q = 10

Hz; dashed line, f_0 = 300 Hz; and dotted line, f_0 = 1000 Hz. Note that the gain response curves of *Figures 8-5, 8-6,* and *8-7* are *normalized* where the maximum response at the center frequency is equal to one, as is the case with the analog response curves of *Figures 8-2* and *8-3*. Note the differences between the response of the analog filter of *Figure 8-2* and the digital filter of *Figure 8-6*: the digital filter response goes to zero (minus infinity on a dB plot) at the Nyquist frequency $f_n = f_s/2$, where in this example, f_s = 11025 Hz and f_n = 5512.5 Hz. If the sample frequency is set to f_s = 44100 Hz or higher, the digital frequency response plots

Second-Order Bandpass Filter

more closely resemble the analog filter responses of *Figure 8-2*. Using the series expansion of the trigonometric functions, cos x, sin x, and tan x:

$$\left.\begin{array}{l} \sin\theta \to \theta \\ \cos\theta \to 1 - \tfrac{1}{2}\theta^2 \\ \tan\theta \to \theta \end{array}\right\} \text{ for } \theta \to 0 \qquad (8\text{-}13)$$

then substituting in Equation (8-11a):

$$G(f) \to \left. \frac{(\theta\,\theta_0/(2Q))}{\sqrt{(\theta\,\theta_0/(2Q))^2 + \left(1 - \tfrac{1}{2}\theta^2 - 1 + \tfrac{1}{2}\theta_0^2\right)^2}} \right\} \text{ for } \theta \to 0$$

$$= \frac{\theta\,\theta_0/(2Q)}{\sqrt{(\theta\,\theta_0/(2Q))^2 + \tfrac{1}{4}\left(-\theta^2 + \theta_0^2\right)^2}} \qquad (8\text{-}14)$$

$$\approx \frac{\theta\,\theta_0/Q}{\sqrt{(\theta\,\theta_0/Q)^2 + \left(\theta_0^2 - \theta^2\right)^2}} = \frac{f\,f_0/Q}{\sqrt{(f\,f_0/Q)^2 + \left(f_0^2 - f^2\right)^2}}$$

Comparing the digital gain response of Equation (8-14) with the analog response of Equation (8-5a) shows that they are identical.

As the sample frequency f_s of an IIR filter is increased, the frequency response approaches that of its analog filter counterpart. Likewise, as the center frequency f_0 and operating frequency f are decreased, the digital filter frequency response approaches that of its analog filter counterpart.

The roll-off of the IIR bandpass on either side of the center frequency is –20 dB/octave, as is the case with the analog bandpass filter. However, a departure from this relationship will occur near the Nyquist frequency, at which point the gain response goes to zero, or minus infinity on a dB plot. This results in a larger attenuation of high frequencies than would be expected from the analog bandpass filter.

DSP Filters

The magnitude of the gain and phase at the center frequency can easily be determined by substituting $\theta = \theta_0$ in Equations (8-12a):

$$G(f_0) = \frac{\sin\theta_0 \, \tan\frac{\theta_0}{2Q}}{\sqrt{\left(\sin\theta_0 \, \tan\frac{\theta_0}{2Q}\right)^2 + (\cos\theta_0 - \cos\theta_0)^2}} = 1 \quad (8\text{-}15a)$$

$$\phi(f_0) = 180\left[\frac{1}{\pi}\tan^{-1}\left(\frac{\cos\theta_0 - \cos\theta_0}{\sin\theta_0 \, \tan\frac{\theta_0}{2Q}}\right)\right] = 0° \quad (8\text{-}15b)$$

Gain and phase response of IIR bandpass filter at its center frequency

Figure 8-7 shows the gain G(f) and phase φ(f) plots for the second order IIR bandpass network of Figure 8-4 for various values of *quality factor Q* with $f_0 = 300$: solid line, Q = 0.2; dotted line, Q = 0.5; thin solid line, Q = 1; thin dashed line, Q = 3; and dashed line, Q = 10.

The quality factor Q is inversely proportional to the frequency bandwidth Δf of the gain response curve:

$$Q \equiv \frac{f_0}{\Delta f} = \frac{\theta_0}{\Delta\theta} \quad (8\text{-}16a)$$

where $\Delta f = f_2 - f_1$ and $\Delta\theta = \theta_2 - \theta_1$. The normalized center frequency θ_0 is related to the normalized edge frequencies θ_1 and θ_2 by:

$$\tan(\theta_0/2) = \sqrt{\tan(\theta_1/2)\tan(\theta_2/2)} \quad (8\text{-}16b)$$

where $\theta_0 = 2\pi f_0 / f_s$, $\theta_1 = 2\pi f_1 / f_s$, and $\theta_2 = 2\pi f_2 / f_s$.

Second-Order Bandpass Filter

The Q of the IIR bandpass filter has a finite limit — not on the high side as one might expect, but rather on the low end:

$$Q > f_0 / f_n \quad \text{or} \quad Q > 2 f_0 / f_s \quad (8\text{-}16c)$$

Q restriction of IIR bandpass filter

There is no theoretical upper limit on Q with the second-order IIR bandpass filter. However, issues concerning round-off errors and other problems with finite size numbers lead to a practical upper limit on Q. This upper limit is determined by word size (number of bits), as well as other factors. For a 24-bit DSP, for example, Qs of 10 to 100 are usually achievable without problems.

The bandwidth Δf of a digital filter must be smaller than the Nyquist frequency $f_n = f_s / 2$.

$$\Delta f < f_n \quad \text{or} \quad \Delta f < f_s / 2$$

The *band-edge* frequencies f_1 and f_2 are defined by the following relationships:

$$f_1 = \frac{f_s}{\pi} \tan^{-1}\left\{ \frac{\tan^2(\pi f_0 / f_s)}{\tan[(f_1 + f_0/Q)\pi / f_s]} \right\} \quad (8\text{-}17a)$$

$$f_2 = \frac{f_s}{\pi} \tan^{-1}\left\{ \frac{\tan^2(\pi f_0 / f_s)}{\tan[(f_2 - f_0/Q)\pi / f_s]} \right\} \quad (8\text{-}17b)$$

where,

$$Q \equiv \frac{f_0}{\Delta f} \qquad \Delta f \equiv f_2 - f_1 \qquad \tan^2(\pi f_0 / f_s) = \tan(\pi f_1 / f_s)\tan(\pi f_2 / f_s)$$

Definition of quality factor Q and its relationship to bandwidth Δf of the IIR bandpass filter

DSP Filters

Note that Equations (8-17) require an iterative solution. However, once this is done, the gain and phase values at the edge frequencies can be found by plugging in the values for f_1 and f_2 from Equations (8-17) into the gain and phase formulas from Equation (8-12):

$$G(f_1) = \frac{\sin\theta_1 \, \tan\frac{\theta_0}{2Q}}{\sqrt{\left(\sin\theta_1 \, \tan\frac{\theta_0}{2Q}\right)^2 + (\cos\theta_1 - \cos\theta_0)^2}} = \frac{1}{\sqrt{2}} \quad (8\text{-}18a)$$

$$G(f_2) = \frac{\sin\theta_2 \, \tan\frac{\theta_0}{2Q}}{\sqrt{\left(\sin\theta_2 \, \tan\frac{\theta_0}{2Q}\right)^2 + (\cos\theta_2 - \cos\theta_0)^2}} = \frac{1}{\sqrt{2}} \quad (8\text{-}18b)$$

$$\phi(f_1) = 180\left[\frac{1}{\pi}\tan^{-1}\left(\frac{\cos\theta_1 - \cos\theta_0}{\sin\theta_1 \, \tan\frac{\theta_0}{2Q}}\right)\right] = 45° \quad (8\text{-}18c)$$

$$\phi(f_2) = 180\left[\frac{1}{\pi}\tan^{-1}\left(\frac{\cos\theta_2 - \cos\theta_0}{\sin\theta_2 \, \tan\frac{\theta_0}{2Q}}\right)\right] = -45° \quad (8\text{-}18d)$$

Gain and response of second-order IIR bandpass filter at its edge frequencies

Difference Equation

The second-order bandpass filter, as with any other digital filter, can be implemented in the time domain using a network such as the one shown in *Figure 8-4* (there are also less common implementations that construct digital filters in the frequency domain). If we start with the analog bandpass transfer function $H(s)$ of Equation (8-3), then apply the bilinear transform:

Second-Order Bandpass Filter

$$s = 2f_s \left(\frac{1-z^{-1}}{1+z^{-1}} \right) \quad (8\text{-}19a)$$

where the analog center frequency Ω_0 is related to the equivalent digital filter center frequency θ_0 by the bilinear *warping* function:

$$\Omega_0 = 2f_s \tan(\theta_0/2) \quad (8\text{-}19b)$$

we arrive at H(z), Equation (8-10). As a warning: *The work involved in this algebraic and trigonometric transformation can be rather formidable (don't try this at home alone).*

Next, using the inverse z-transform techniques presented in Chapter 3, Equation (8-10) can be converted into a difference equation, which is then implemented in software (or DSP hardware):

$$y(n) = 2\{\alpha\,[x(n) - x(n-2)] + \gamma\, y(n-1) - \beta\, y(n-2)\} \quad (8\text{-}20a)$$

$$\beta = \frac{1}{2}\frac{1-\tan[\theta_0/(2Q)]}{1+\tan[\theta_0/(2Q)]} \qquad \gamma = (\tfrac{1}{2}+\beta)\cos\theta_0 \qquad \alpha = (\tfrac{1}{2}-\beta)/2 \quad (8\text{-}20b)$$

$$\text{where, } \theta \equiv 2\pi f / f_s \qquad \theta_0 \equiv 2\pi f_0 / f_s$$

Difference equation of the digital second-order bandpass filter with coefficient formulas

where x(n) is the current input sample; x(n-1) is the previous input; x(n-2) is the previous previous input; y(n-1) is the previous output; y(n-2) is the previous previous output; and y(n) is the current output.

Below is an example C language implementation of the second-order bandpass filter difference equation:

DSP Filters

```cpp
//
// cook_bandpass.cpp
//
// Implementation of a simple Bandpass Filter Stage,
// class CBandpassFilterStage

#include "cook.h"

CBandpassFilterStage::CBandpassFilterStage()
{
      x1 = 0;
      x2 = 0;
      y1 = 0;
      y2 = 0;
}


CBandpassFilterStage::~CBandpassFilterStage()
{
}

void CBandpassFilterStage::execute_filter_stage()
{
      y = 2 * ((alpha * (x - x2) + gamma * y1 - beta * y2);
      x2 = x1;
      x1 = x;
      y2 = y1;
      y1 = y;
}
```

9
SECOND-ORDER BAND-STOP FILTER

Analog Filter Network

A band-stop filter can be implemented using the RCL network shown in *Figure 9-1*, where the impedance values of the individual components are given by:

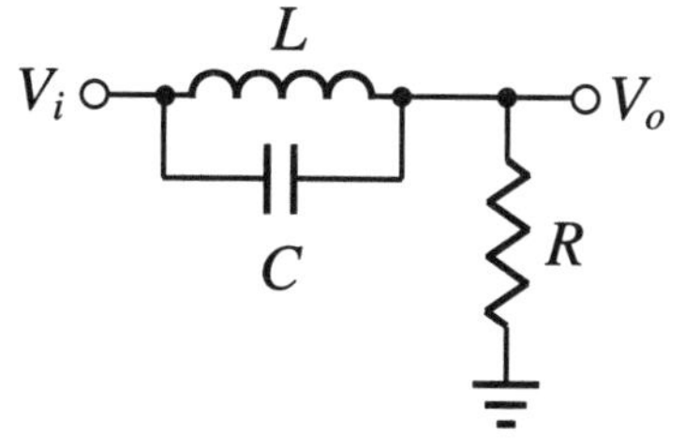

Figure 9-1. RCL band-stop filter network

$$\Omega_0 \equiv \frac{1}{\sqrt{LC}}$$

$$Q \equiv R\sqrt{\frac{C}{L}}$$

$$Z_R = R \qquad Z_C = \frac{1}{sC} \qquad Z_L = sL \qquad (9\text{-}1)$$

DSP Filters

The Laplace variable s is defined by $s = j\Omega = 2\pi j f$, where $j \equiv \sqrt{-1}$ (f is frequency in Hertz). The set of units used in our calculations are derived from the *mks* (meters-kilograms-seconds) system of measure, which is to say that ohms are used for R, farads for C, and henrys for L.

The filter of *Figure 9-1* is expressed as the ratio of the output V_o to input V_i and is equivalent to the ratio of the series resistor to the total resistance of the network:

$$\frac{V_o}{V_i} = \frac{Z_R}{Z_L \| Z_C + Z_R} \qquad (9\text{-}2a)$$

Combining the expressions for complex impedance from Equations (9-1) for the R, C, and L components, results in a filter *transfer function* $H(s)$:

$$H(s) \equiv \frac{R(sL + 1/sC)}{L/C + R(sL + 1/sC)} \qquad (9\text{-}2b)$$

where $H(s)$ is defined as the ratio of V_o to input V_i. $H(s)$ can be algebraically manipulated by dividing both the numerator and denominator by LR and multiplying by s:

$$H(s) = \frac{s^2 + 1/LC}{s^2 + s/RC + 1/LC}$$

$$= \frac{s^2 + 1/LC}{s^2 + s \dfrac{1}{LC}\sqrt{\dfrac{L}{R^2 C}} + \dfrac{1}{LC}} \qquad (9\text{-}2c)$$

Simplifying Equation (9-2c) results in the following analog band-stop filter transfer function:

$$H(s) = \frac{s^2 + \Omega_0^2}{s^2 + s\Omega_0/Q + \Omega_0^2} \qquad (9\text{-}3)$$

where,

$$\Omega_0 \equiv \frac{1}{\sqrt{LC}} \qquad Q \equiv R\sqrt{\frac{C}{L}} \qquad (9\text{-}4)$$

Second-Order Band-Stop Filter

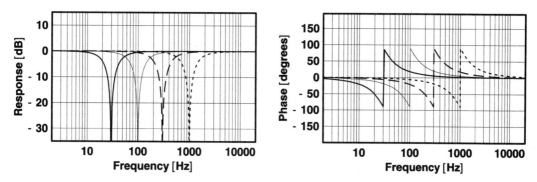

Figure 9-2. Gain response and phase plots for analog band-stop filter with Q = 1: solid line, $f_0 = 30$; thin solid line, $f_0 = 100$; dashed line, $f_0 = 300$; dotted line, $f_0 = 1000$

where the damping factor $d = 1/Q$. The magnitude $G(f)$ and the phase angle $\phi(f)$ of $H(s)$ are summarized in Equations (9-5):

$$G(f) = \frac{|f^2 - f_0^2|}{\sqrt{(f_0^2 - f^2)^2 + (f_0 f / Q)^2}} \qquad \phi(f) = -\frac{180}{\pi} \tan^{-1}\left(\frac{f_0 f / Q}{f_0^2 - f^2}\right) \qquad (9\text{-}5a)$$

$$f_0 \equiv \frac{1}{2\pi \sqrt{LC}} \qquad Q \equiv R\sqrt{\frac{C}{L}} \qquad (9\text{-}5b)$$

Magnitude and phase response of the analog second-order band-stop filter

Figure 9-2 shows examples of gain and phase response curves for the analog second-order band-stop filter using several values of center frequency, with $Q = 1$: thick solid line, $f_0 = 30$ Hz; thin solid line, $f_0 = 100$ Hz; dashed line, $f_0 = 300$ Hz; and dotted line, $f_0 = 1000$ Hz.

The magnitude and phase response values at the center frequency can conveniently be determined by substituting $f = f_0$ in Equations (9-5a):

DSP Filters

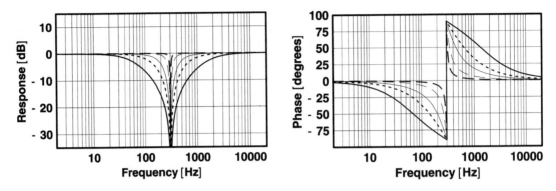

Figure 9-3. Gain response and phase plots for analog band-stop filter with $f_0 = 300$: solid line, $Q = 0.2$; dotted line, $Q = 0.5$; thin solid line, $Q = 1$; thin dashed line, $Q = 3$; dashed line, $Q = 10$

$$G(f_0) = \frac{|f_0^2 - f_0^2|}{\sqrt{(f_0^2 - f_0^2)^2 + (f_0 f_0 / Q)^2}} = \boxed{0} \quad (9\text{-}6a)$$

$$\phi(f_0) = -180\left[\frac{1}{\pi}\tan^{-1}\left(\frac{f_0 f_0 / Q}{f_0^2 - f_0^2}\right)\right] = \boxed{\text{undefined}} \quad (9\text{-}6b)$$

Gain and phase response of analog band-stop filter at its center frequency

Figure 9-3 shows the gain $G(f)$ and phase $\phi(f)$ plots for the second order band-stop RCL network of Figure 9-1 for various values of *quality factor Q* with $f_0 = 300$: solid line, $Q = 0.2$; dotted line, $Q = 0.5$; thin solid line, $Q = 1$; thin dashed line, $Q = 3$; and dashed line, $Q = 10$.

Second-Order Band-Stop Filter

As discussed in the previous chapter, the quality factor Q is inversely proportional to the frequency bandwidth Δf of the gain response curve:

$$Q \equiv \frac{f_0}{\Delta f} \quad (9\text{-}7a)$$

where $\Delta f = f_2 - f_1$. The center frequency f_0 is equal to the geometric mean of f_1 and f_2:

$$f_0 = \sqrt{f_1 f_2} \quad (9\text{-}7b)$$

As in the case of the bandpass filter, the *band-edge* frequencies f_1 and f_2 are defined by the following relationships:

$$f_1 = \frac{f_0}{2Q}\left(\sqrt{1 + 4Q^2} - 1\right) \quad (9\text{-}8a)$$

$$f_2 = \frac{f_0}{2Q}\left(\sqrt{1 + 4Q^2} + 1\right) \quad (9\text{-}8b)$$

where,

$$Q \equiv \frac{f_0}{\Delta f} \qquad \Delta f \equiv f_2 - f_1 \qquad f_0 = \sqrt{f_1 f_2}$$

Definition of quality factor Q and its relationship to bandwidth Δf

DSP Filters

Following the same procedure as described in Chapter 8, the gain and phase values at the edge frequencies can be found by plugging in the expressions for f_1 and f_2 from Equations (9-8) into the formulas from Equation (9-5):

$$G(f_1) = \frac{|f_1^2 - f_0^2|}{\sqrt{(f_0^2 - f_1^2)^2 + (f_0 f_1 / Q)^2}} = \boxed{\frac{1}{\sqrt{2}}} \qquad (9\text{-}9a)$$

$$G(f_2) = \frac{|f_2^2 - f_0^2|}{\sqrt{(f_0^2 - f_2^2)^2 + (f_0 f_2 / Q)^2}} = \boxed{\frac{1}{\sqrt{2}}} \qquad (9\text{-}9b)$$

$$\phi(f_1) = -180 \left[\frac{1}{\pi} \tan^{-1}\left(\frac{f_0 f_1 / Q}{f_0^2 - f_1^2} \right) \right] = \boxed{-45°} \qquad (9\text{-}9c)$$

$$\phi(f_2) = -180 \left[\frac{1}{\pi} \tan^{-1}\left(\frac{f_0 f_2 / Q}{f_0^2 - f_2^2} \right) \right] = \boxed{45°} \qquad (9\text{-}9d)$$

Gain and phase response of analog band-stop filter at the edge frequencies

Digital Filter Network

A second-order IIR band-stop filter is constructed by implementing the network of Figure 9-4 on a stream of sampled data, $x(n)$; using two

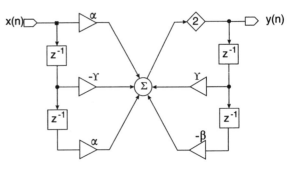

Figure 9-4. Digital band-stop filter network

Second-Order Band-Stop Filter

previous *input states*, *x(n-1)* and *x(n-2)*; and two previous *output states*, *y(n-1)* and *y(n-2)*. At every sample time, $nT = n/f_s$, one output value *y(n)* is generated. This sample-by-sample processing is common in many filter applications. The basic components of the digital filter network shown in *Figure 9-4* are the *accumulator* and *delay* element. The accumulator is made up of a multiplier (triangle) and adder (circle). When using a general-purpose digital signal processor (DSP), these components correspond to key instructions of the DSP's instruction set.

Like the s-domain transfer function of Equation (9-3), the transfer function of the band-stop network of *Figure 9-4* describes its response characteristics, but is instead a function of the z-transform:

$$H(z) = \frac{\alpha - \gamma z^{-1} + \alpha z^{-2}}{\frac{1}{2} - \gamma z^{-1} + \beta z^{-2}} \quad (9\text{-}10)$$

One useful means of quantifying the relationship between s and z can be specified by the bilinear transformation (see previous chapter). Recall from these previous discussions that the s variable of the Laplace transform represents a point in the complex plane, such that the analog filter response characteristics are found along the vertical line defined by $s = j2\pi f$. In a similar manner, the z variable of the z-transform is again a point in the complex plane, but the digital filter response is found along the unit circle at $z = e^{j2\pi f/f_s}$, where f_s is the sample frequency.

The coefficients α, β, and γ in Equation (9-10) are directly related to the filter quality factor Q and center frequency f_o, by:

$$\beta = \frac{1}{2}\left(\frac{1 - \tan\frac{\theta_0}{2Q}}{1 + \tan\frac{\theta_0}{2Q}}\right)$$

$$\gamma = \left(\tfrac{1}{2} + \beta\right)\cos(\theta_0) \quad (9\text{-}11)$$

$$\alpha = \left(\tfrac{1}{2} + \beta\right)/2$$

DSP Filters

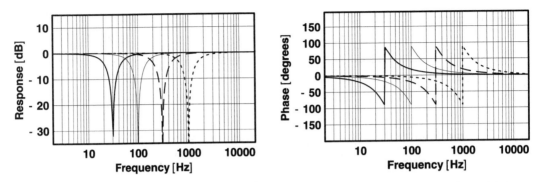

Figure 9-5. Gain and phase of digital band-stop filter with Q = 1 and f_s = 44100: solid line, f_0 = 30; thin solid line, f_0 = 100; dashed line, f_0 = 300; dotted line, f_0 = 1000

where $\theta_0 = 2\pi f_0 / f_s$, which is the *normalized center frequency*. The digital band-stop filter frequency response is the magnitude G(f) and the phase angle ɸ(f) of H(z):

$$G(f) = \frac{|\cos\theta - \cos\theta_0|}{\sqrt{\left(\sin\theta \ \tan\frac{\theta_0}{2Q}\right)^2 + (\cos\theta - \cos\theta_0)^2}} \quad (9\text{-}12a)$$

$$\phi(f) = -\frac{180}{\pi} \tan^{-1}\left(\frac{\sin\theta \ \tan\frac{\theta_0}{2Q}}{\cos\theta - \cos\theta_0}\right) \quad (9\text{-}12b)$$

$$\text{where,} \quad \theta \equiv 2\pi f / f_s \qquad \theta_0 \equiv 2\pi f_0 / f_s$$

Magnitude and phase response of the digital second-order band-stop filter

Figure 9-5 shows examples of magnitude and phase responses for the digital second-order band-stop filter, using various values of center frequency, with Q = 1: thick solid line, f_0 = 30 Hz; thin solid line, f_0 = 100

Second-Order Band-Stop Filter

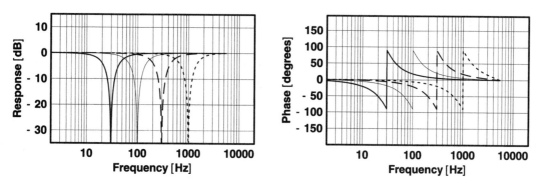

Figure 9-6. Gain and phase of digital band-stop filter with Q = 1 and f_s = 11025: solid line, f_0 = 30; thin solid line, f_0 = 100; dashed line, f_0 = 300; and dotted line, f_0 = 1000

Hz; dashed line, f_0 = 300 Hz; and dotted line, f_0 = 1000 Hz. Note that the gain response curves of *Figures 9-5, 9-6,* and *9-7* are normalized where the maximum response away from the center frequency is equal to one, as is the case with the analog response curves of *Figures 9-2* and *9-3.* There is no noticeable difference between the response of the analog filter of *Figure 9-2* and the digital filter of *Figure 9-5.*

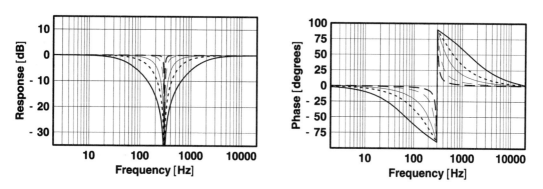

Figure 9-7. Gain response and phase plots for IIR band-stop filter with f_s = 44100 and f_0 = 300: solid line, Q = 0.2; dotted line, Q = 0.5; thin solid line, Q = 1; thin dashed line, Q = 3; dashed line, Q = 10

DSP Filters

Using the series expansion of the trigonometric functions, $\cos x$, $\sin x$, and $\tan x$:

$$\left.\begin{array}{c} \sin\theta \to \theta \\ \cos\theta \to 1-\tfrac{1}{2}\theta^2 \\ \tan\theta \to \theta \end{array}\right\} \text{ for } \theta \to 0 \quad (9\text{-}13)$$

then substituting in Equation (9-11a):

$$G(f) \to \frac{\left|1-\tfrac{1}{2}\theta^2 - 1 + \tfrac{1}{2}\theta_0^2\right|}{\sqrt{(\theta\theta_0/(2Q))^2 + \left(1-\tfrac{1}{2}\theta^2 - 1 + \tfrac{1}{2}\theta_0^2\right)^2}} \quad \text{for } \theta \to 0$$

$$= \frac{\left|-\tfrac{1}{2}\theta^2 + \tfrac{1}{2}\theta_0^2\right|}{\sqrt{(\theta\theta_0/(2Q))^2 + \tfrac{1}{4}\left(-\theta^2 + \theta_0^2\right)^2}} \quad (9\text{-}14)$$

$$= \frac{\left|\theta^2 - \theta_0^2\right|}{\sqrt{(\theta\theta_0/Q)^2 + (\theta_0^2 - \theta^2)^2}} = \frac{\left|f^2 - f_0^2\right|}{\sqrt{(f f_0/Q)^2 + (f_0^2 - f^2)^2}}$$

The result of Equation (9-14) is identical in form to the analog gain response of Equation (9-5a). A similar result is obtained by substituting Equation (9-13) in the IIR phase formula of Equation (9-12b). It will again be identical to the analog band-stop phase response formula of Equation (9-5b).

The magnitude of the gain and phase at the center frequency can easily be determined by substituting $\theta = \theta_0$ in Equations (9-12a):

Second-Order Band-Stop Filter

$$G(f_0) = \frac{|\cos\theta_0 - \cos\theta|}{\sqrt{\left(\sin\theta_0 \tan\frac{\theta_0}{2Q}\right)^2 + (\cos\theta_0 - \cos\theta_0)^2}} = 0 \qquad (9\text{-}15a)$$

$$\phi(f_0) = -\frac{180}{\pi}\tan^{-1}\left(\frac{\sin\theta_0 \tan\frac{\theta_0}{2Q}}{\cos\theta_0 - \cos\theta_0}\right) = \text{undefined} \qquad (9\text{-}15b)$$

Gain and phase response of IIR band-stop filter at its center frequency

Figure 9-7 shows the gain G(f) and phase φ(f) plots for the second order IIR band-stop network of Figure 9-4, for various values of quality factor Q with f_0 = 300: solid line, Q = 0.2; dotted line, Q = 0.5; thin solid line, Q = 1; thin dashed line, Q = 3; and dashed line, Q = 10.

The quality factor Q is inversely proportional to the frequency bandwidth Δf of the gain response curve:

$$Q \equiv \frac{f_0}{\Delta f} = \frac{\theta_0}{\Delta\theta} \qquad (9\text{-}16a)$$

where $\Delta f = f_2 - f_1$ and $\Delta\theta = \theta_2 - \theta_1$. The normalized center frequency θ_0 is related to the normalized edge frequencies θ_1 and θ_2 by:

$$\tan(\theta_0/2) = \sqrt{\tan(\theta_1/2)\tan(\theta_2/2)} \qquad (9\text{-}16b)$$

where $\theta_0 = 2\pi f_0/f_s$, $\theta_1 = 2\pi f_1/f_s$, and $\theta_2 = 2\pi f_2/f_s$.

The Q of the IIR band-stop filter has a lower limit, which is identical to that of the bandpass filter:

DSP Filters

$$Q > f_0 / f_n \quad \text{or} \quad Q > 2 f_0 / f_s \quad (9\text{-}16c)$$

Q restriction of IIR band-stop filter

As with the bandpass filter, there is no theoretical upper limit on Q of the second order IIR band-stop filter. However, round-off errors and other problems with finite word size lead to an upper limit determined mostly by the number of bits per word (numerical precision), as well as other factors such as the ratio of the center frequency to sample frequency.

The bandwidth Δf of a digital filter must be smaller than the Nyquist frequency, $f_n = f_s / 2$.

$$\Delta f < f_n \quad \text{or} \quad \Delta f < f_s / 2$$

The band-edge frequencies f_1 and f_2 are defined by the following relationships:

$$f_1 = \frac{f_s}{\pi} \tan^{-1}\left\{ \frac{\tan^2(\pi f_0 / f_s)}{\tan[(f_1 + f_0 / Q)\pi / f_s]} \right\} \quad (9\text{-}17a)$$

$$f_2 = \frac{f_s}{\pi} \tan^{-1}\left\{ \frac{\tan^2(\pi f_0 / f_s)}{\tan[(f_2 + f_0 / Q)\pi / f_s]} \right\} \quad (9\text{-}17b)$$

where,

$$Q \equiv \frac{f_0}{\Delta f} \qquad \Delta f \equiv f_2 - f_1 \qquad \tan^2(\pi f_0 / f_s) = \tan(\pi f_1 / f_s)\tan(\pi f_2 / f_s)$$

Definition of quality factor Q and its relationship to bandwidth Δf of the IIR band-stop filter

Second-Order Band-Stop Filter

As in the case of the bandpass filter of Chapter 8, determining the edge frequencies from Equations (9-17) requires an iterative solution. Substituting the final estimates of f_1 and f_2 into the band-stop gain and phase formulas of Equation (9-12), result in the gain and phase values at the edge frequencies:

$$G(f_1) = \frac{|\cos\theta_1 - \cos\theta_0|}{\sqrt{\left(\sin\theta_1 \ \tan\frac{\theta_0}{2Q}\right)^2 + (\cos\theta_1 - \cos\theta_0)^2}} = \frac{1}{\sqrt{2}} \quad (9\text{-}18a)$$

$$G(f_2) = \frac{|\cos\theta_2 - \cos\theta_0|}{\sqrt{\left(\sin\theta_2 \ \tan\frac{\theta_0}{2Q}\right)^2 + (\cos\theta_2 - \cos\theta_0)^2}} = \frac{1}{\sqrt{2}} \quad (9\text{-}18b)$$

$$\phi(f_1) = -\frac{180}{\pi}\tan^{-1}\left(\frac{\sin\theta_1 \ \tan\frac{\theta_0}{2Q}}{\cos\theta_1 - \cos\theta_0}\right) = -45° \quad (9\text{-}18c)$$

$$\phi(f_2) = -\frac{180}{\pi}\tan^{-1}\left(\frac{\sin\theta_2 \ \tan\frac{\theta_0}{2Q}}{\cos\theta_2 - \cos\theta_0}\right) = 45° \quad (9\text{-}18d)$$

Gain and phase response of second-order IIR band-stop filter at its edge frequencies

Difference Equation

Equation (9-19a) is the *direct form* difference equation for the second order band-stop filter, where the input, x(n); input states, x(n) and x(n-1); output states, y(n-1) and y(n-2); and filter coefficients, α, β, and γ compute a *sum of products*, resulting in the output sample y(n). The numerical processing of *filter states* and *filter coefficients* is illustrated by the band-stop network diagram of *Figure 9-4*.

Example C code which implements the band-stop difference equation is shown on the following page:

105

DSP Filters

$$y(n) = 2\{\alpha\, x(n) - \gamma\, x(n-2) + \alpha\, x(n-2) + \gamma\, y(n-1) - \beta\, y(n-2)\} \quad (9\text{-}19a)$$

$$\beta = \frac{1}{2}\frac{1-\tan[\theta_0/(2Q)]}{1+\tan[\theta_0/(2Q)]} \qquad \gamma = \left(\tfrac{1}{2}+\beta\right)\cos\theta_0 \qquad \alpha = \left(\tfrac{1}{2}+\beta\right)/2 \quad (9\text{-}19b)$$

$$\text{where,} \quad \theta \equiv 2\pi f/f_s \qquad \theta_0 \equiv 2\pi f_0/f_s$$

Difference equation of the digital second-order band-stop filter with coefficient formulas

```
//
// cook_bandstop.cpp
//
// Implementation of a simple Bandstop Filter Stage,
// class CBandstopFilterStage

#include "cook.h"

CBandstopFilterStage::CBandstopFilterStage()
{
        x1 = 0;
        x2 = 0;
        y1 = 0;
        y2 = 0;
}

CBandstopFilterStage::~CBandstopFilterStage()
{
}

void CBandstopFilterStage::execute_filter_stage()
{
        y = 2*(alpha*x - gamma*x1 + alpha*x2 + gamma*y1 - beta*y2);
        x2 = x1;
        x1 = x;
        y2 = y1;
        y1 = y;
}
```

10

PEAKING FILTER

Digital Filter Network

A second-order IIR *peaking* filter can be implemented by summing the input $x(n)$ with the output of a second-order bandpass filter, scaled by μ-1, as shown in *Figure 10-1*. The bandpass output scale factor is chosen so that when $\mu = 1$, the output is equal to the input, $y(n) = x(n)$. The coefficient depends on the peaking level g — as $\mu \equiv 10^{g/20}$ — where g is the boost/cut gain in dB.

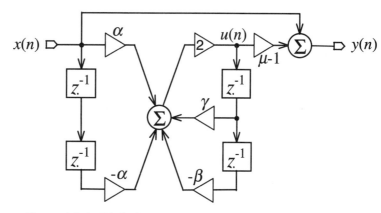

Figure 10-1. Digital peaking filter network

The output of the second-order peaking filter is generated by computing a convolution (running average) on a stream of sampled input

DSP Filters

data using two previous inputs and two previous outputs for each new input and output value. Because of the use of previous outputs, this type of filter is called a recursive filter or an infinite impulse response (IIR). Unlike second-order IIR filters discussed previously, the peaking filter has two *summing junctions*, as can be seen by examining *Figure 10-1*. The left summing junction belongs to the bandpass filter subsection, which is described in detail in Chapter 8. The right summing junction forms a first-order *finite impulse response* (FIR) filter subsection by combining the input x(n) with the weighted value of u(n). The total network is second order because the filter states making up the network only go back in time (discrete time) by two samples: x(n-1) is the previous input; x(n-2) is the previous previous input; u(n-1) is the previous output of the bandpass filter subsection; and u(n-2) is the previous previous output of the bandpass section.

According to the network diagram of *Figure 10-1*, the filter transfer function for the total network can be expressed as:

$$H(z) = 1 + (\mu - 1) \frac{\alpha(1 - z^{-2})}{\frac{1}{2} - \gamma z^{-1} + \beta z^{-2}} \quad (10\text{-}1)$$

The coefficients α, β, and γ in Equation (10-1) are directly related to the bandpass filter section quality factor Q, center frequency f_0, and the peaking filter boost/cut gain g by:

$$\beta = \frac{1}{2} \left(\frac{1 - \left(\frac{4}{1+\mu}\right) \tan \frac{\theta_0}{2Q}}{1 + \left(\frac{4}{1+\mu}\right) \tan \frac{\theta_0}{2Q}} \right)$$

$$\gamma = \left(\tfrac{1}{2} + \beta\right) \cos(\theta_0)$$

$$\alpha = \left(\tfrac{1}{2} - \beta\right) / 2$$

(10-2)

where $\theta_0 = 2\pi f_0 / f_s$, which is the normalized center frequency; and $\mu = 10^{g/20}$, where again, g is the boost/cut gain in dB.

Peaking Filter

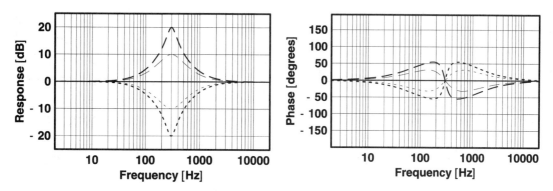

Figure 10-2. Gain and phase of digital peaking filter with Q = 1, f_0 = 300, and fs = 44100: dotted line, g = -20 dB; thin dotted line, g = -10dB; solid line, g = 0 dB; thin dashed line, g = +10 db; dashed line, g = +20 dB

The IIR peaking filter Q has a finite limit due to the lower Q limit of the bandpass filter subsection:

$$Q > f_0/f_n$$

Q restriction of IIR peaking filter

As with the bandpass filter, the upper Q limit is determined by numerical precision of the number base in use.

Figure 10-2 shows several examples of gain and phase response curves for the digital second-order peaking filter, using several values of boost/cut gain g with Q = 1, f_0 = 300 Hz, and f_s = 44100 Hz: dotted line, g = –20 dB; thin dotted line, g = –10 dB; solid line, g = 0 dB; thin dashed line, g = +10 dB; and dashed line, g = +20 dB. Note that since the network is second order, the phase excursion will not exceed ±90°. The maximum and minimum phase value is controlled by the peaking filter boost/cut gain factor g. With the specific choice of boost/cut gains used in Figure 10-2, the phase does not exceed approximately ±55° at g = ±20 dB.

109

DSP Filters

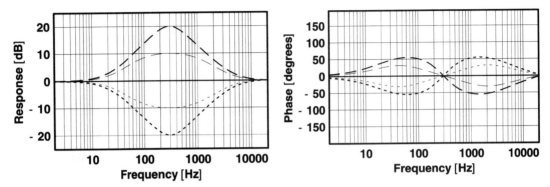

Figure 10-3. Gain and phase of digital peaking filter with $Q = 0.25$, $f_0 = 300$, and $f_s = 44100$: dotted line, $g = -20$ dB; thin dotted line, $g = -10$ dB; solid line, $g = 0$ dB; thin dashed line, $g = +10$ dB; dashed line, $g = +20$ dB

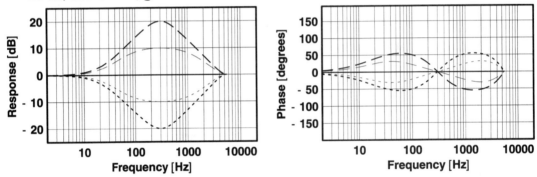

Figure 10-4. Gain and phase of digital peaking filter with $Q = 0.25$, $f_0 = 300$, and $f_s = 11025$: dotted line, $g = -20$ dB; thin dotted line, $g = -10$ dB; solid line, $g = 0$ dB; thin dashed line, $g = +10$ dB; dashed line, $g = +20$ dB

Figure 10-3 is identical to *Figure 10-2*, with the exception that $Q = 0.25$, and *Figure 10-4* is identical to *Figure 10-3*, with the exception that $f_s = 11025$. *Figure 10-4* clearly demonstrates that as the sample frequency is decreased, the response curves converge to 0 dB and 0° at the Nyquist frequency. This characteristic is similar to the behavior of the bandpass filter of Chapter 8 and the low-pass filter of Chapter 6.

Figure 10-5 shows several examples of gain and phase response curves for the IIR peaking filter, using several values of center frequency with a boost/cut gain $g = +15$ dB, $Q = 1$, and sample frequency $f_s =$

Peaking Filter

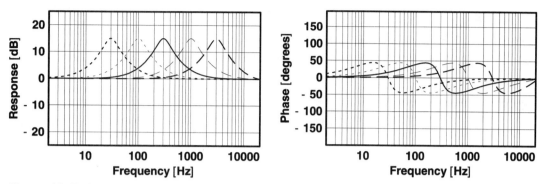

Figure 10-5. Gain and phase of digital peaking filter with Q = 1, f_s = 44100, and g = +15 dB: dotted line, f_0 = 30; thin dotted line, f_0 = 100; solid line, f_0 = 300; thin dashed line, f_0 = 1000; and dashed line, f_0 = 3000 Hz

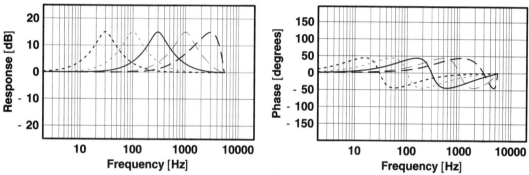

Figure 10-6. Gain and phase of digital peaking filter with Q = 1, f_s = 11025 and g = +15 dB: dotted line, f_0 = 30; thin dotted line, f_0 = 100; solid line, f_0 = 300; thin dashed line, f_0 = 1000; and dashed line, f_0 = 3000 Hz

44100 Hz: dotted line, f_0 = 30 Hz; thin dotted line, f_0 = 100 Hz; solid line, f_0 = 300 Hz; thin dashed line, f_0 = 1000 Hz; and dashed line, f_0 = 3000 Hz. Note that with the specific choice of boost/cut gains used in *Figure 10-5*, the phase does not exceed approximately ±45°.

Figure 10-6 is identical to *Figure 10-5* with the exception that f_s = 11025. *Figure 10-6* again demonstrates that as the sample frequency is decreased, the response curves converge to 0 dB and 0° at the Nyquist frequency. This characteristic has the effect of warping the response curves from the normal symmetric shape that is seen at low operating frequencies and/or high sample frequencies.

DSP Filters

The gain and phase response of the second-order peaking filter can be calculated by using the response formulas of the second-order bandpass filter from Chapter 8 combined with the FIR subsection of the total peaking filter network:

$$G(f) = \sqrt{X^2(f)+Y^2(f)} \qquad \phi(f) = \frac{180}{\pi}\tan^{-1}\frac{Y(f)}{X(f)} \qquad (10\text{-}3a)$$

$$X(f) \equiv 1 + (\mu-1)\cos(\phi_B(f))G_B(f) \qquad (10\text{-}3b)$$

$$Y(f) \equiv (\mu-1)\sin(\phi_B(f))G_B(f) \qquad (10\text{-}3c)$$

where,

$$G_B(f) \equiv \frac{\sin\theta\,\tan\dfrac{\theta_0}{2Q'}}{\sqrt{\left(\sin\theta\,\tan\dfrac{\theta_0}{2Q'}\right)^2 + (\cos\theta - \cos\theta_0)^2}} \qquad \phi_B(f) \equiv \tan^{-1}\left(\frac{\cos\theta - \cos\theta_0}{\sin\theta\,\tan\dfrac{\theta_0}{2Q'}}\right)$$

$$Q' = \frac{\theta_0}{2}\tan^{-1}\left(\frac{4}{1+\mu}\tan\frac{\theta_0}{2Q}\right)$$

$$\theta \equiv 2\pi f / f_s \qquad \theta_0 \equiv 2\pi f_0 / f_s$$

Magnitude and phase response of the second-order peaking filter

The magnitude of the gain and phase at the center frequency can easily be determined by substituting $f = f_0$ in Equations (10-3):

$$G(f_0) = \sqrt{X^2(f_0)+Y^2(f_0)} \;=\; \sqrt{[1+(\mu-1)]^2 + 0^2} \;=\; \mu \qquad (10\text{-}4a)$$

$$\phi(f_0) = \frac{180}{\pi}\tan^{-1}\frac{Y\{f_0\}}{X\{f_0\}} = \frac{180}{\pi}\tan^{-1}\frac{0}{\mu} = 0° \qquad (10\text{-}4b)$$

where,

$$X(f_0) = 1 + (\mu-1)\cos(\phi_B(f_0))G_B(f_0)$$

$$= 1 + (\mu-1) = \mu \qquad\qquad G_B(f_0) = 1$$

and,

$$Y(f_0) = (\mu-1)\sin(\phi_B(f_0))G_B(f_0) \qquad\qquad \phi_B(f_0) = 0$$

$$= 0$$

Gain and phase response of IIR peaking filter at its center frequency

Peaking Filter

Difference Equation

In order to implement the digital IIR filter, a difference equation is used. As previously mentioned, the IIR filter implementation is simply a running average of the input and output data:

$$u(n) = 2\{\alpha[x(n) - x(n-2)] + \gamma u(n-1) - \beta u(n-2)\} \quad (10\text{-}5a)$$

$$y(n) = x(n) + (\mu - 1)u(n) \quad (10\text{-}5b)$$

where $x(n)$ is the current input sample; $x(n\text{-}1)$ is the previous input; $x(n\text{-}2)$ is the previous previous input; $u(n\text{-}1)$ is the previous output of the bandpass filter section; $u(n\text{-}2)$ is the previous previous output of the bandpass filter section; and $y(n)$ is the current filter output. The filter coefficients α, β, and γ are defined in Equation (10-2), as well as in (10-6c).

$$u(n) = 2\{\alpha[x(n) - x(n-2)] + \gamma u(n-1) - \beta u(n-2)\} \quad (10\text{-}6a)$$

$$y(n) = x(n) + (\mu - 1)u(n) \quad (10\text{-}6b)$$

$$\beta = \frac{1}{2}\left(\frac{1 - \left(\frac{4}{1+\mu}\right)\tan\frac{\theta_0}{2Q}}{1 + \left(\frac{4}{1+\mu}\right)\tan\frac{\theta_0}{2Q}}\right) \qquad \gamma = \left(\tfrac{1}{2} + \beta\right)\cos\theta_0 \qquad \alpha = \left(\tfrac{1}{2} - \beta\right)/2 \quad (10\text{-}6c)$$

$$\text{where,} \quad \theta \equiv 2\pi f / f_s \qquad \theta_0 \equiv 2\pi f_0 / f_s$$

Difference equation of the digital second-order peaking filter with coefficient formulas

DSP Filters

An example C code for the peaking filter:

```
//
// cook_peaking.cpp
//
// Implementation of a simple Peaking Filter Stage,
// class CPeakingFilterStage

#include "cook.h"

CPeakingFilterStage::CPeakingFilterStage()
{
        x1 = 0;
        x2 = 0;
        y1 = 0;
        y2 = 0;
}

CPeakingFilterStage::~CPeakingFilterStage()
{
}


void CPeakingFilterStage::execute_filter_stage()
{
        y = 2 * (alpha * (x - x2) + gamma * y1 - beta * y2);
        x2 = x1;
        x1 = x;
        y2 = y1;
        y1 = y;
        y = (y * (mu - 1.0)) + x;
}
```

11

SHELVING FILTER

Low-Pass IIR

Digital Filter Network

A first-order IIR low-pass *shelving* filter can be implemented by summing the input $x(n)$ with the output of a first-order low-pass filter, scaled by μ-1, as shown in *Figure 11-1*. The low-pass output scale factor is chosen so that when $\mu = 1$ the output is equal to the input, $y(n) = x(n)$. The coefficient depends on the shelving level g, as $\mu \equiv 10^{g/20}$, where g is the boost/cut gain in dB.

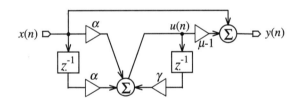

Figure 11-1. IIR low-pass shelving filter

The output of the first-order shelving filter is generated by computing a running average on a stream of sampled input data using one previous input and one previous output for each new input and output value. Because of the use of previous outputs, this type of filter is called a recursive filter or an infinite impulse response (IIR). Unlike the first-order IIR filters already discussed, the shelving filter has two *summing junctions* as can be seen by examining *Figure 11-1*. The left summing junction belongs to the low-pass filter subsection, which is described in detail in Chapter 4. The right summing junction forms a first-order finite impulse response (FIR) filter subsection by combining the input $x(n)$ with the weighted value of $u(n)$. The total network is first order because the filter states making up the network only go back in time (discrete time) by one sample: $x(n-1)$ is the previous input, and $u(n-1)$ is the previous output of the low-pass filter subsection.

According to the network diagram of *Figure 11-1*, the filter transfer function for the total network can be expressed as:

$$H(z) = 1 + (\mu - 1)\frac{\alpha(1+z^{-1})}{1-\gamma z^{-1}} \quad (11\text{-}1)$$

The coefficients α and γ in Equation (11-1) are directly related to the low-pass filter center frequency f_c and the shelving filter boost/cut gain g by:

$$\gamma = \frac{1 - \left(\dfrac{4}{1+\mu}\right)\tan\dfrac{\theta_c}{2}}{1 + \left(\dfrac{4}{1+\mu}\right)\tan\dfrac{\theta_c}{2}} \quad (11\text{-}2)$$

$$\alpha = (1-\gamma)/2$$

where $\theta_c = 2\pi f_c / f_s$, which is the *normalized cutoff frequency*; and $\mu = 10^{g/20}$, where again, g is the boost/cut gain in dB.

Figure 11-2 shows several examples of gain and phase response curves for the first-order low-pass shelving filter using several values of boost/cut gain g with f_c = 30 and f_s = 44100 Hz: dotted line, g = -15 dB;

Shelving Filter

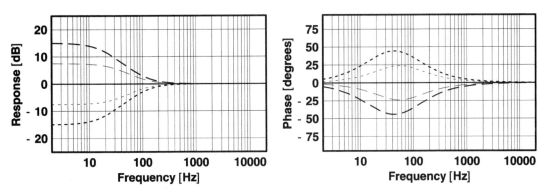

Figure 11-2. Gain and phase of low-pass shelving filter with $f_c = 30$ and $f_s = 44100$: dotted line, $g = -15$ dB; thin dotted line, $g = -7.5$ dB; solid line, $g = 0$ dB; thin dashed line, $g = +7.5$ dB; and dashed line, $g = +15$ dB.

thin dotted line, $g = -7.5$ dB; solid line, $g = 0$ dB; thin dashed line, $g = +7.5$ dB; and dashed line, $g = +15$ dB. Note that since the network is first order, the phase excursion will not exceed ±90°. The maximum and minimum phase value is controlled by the shelving filter boost/cut gain factor g. With the specific choice of boost/cut gains used in *Figure 11-2*, the phase does not exceed approximately ±45° at $g = ±15$ dB. *Figure 11-3* shows several examples of gain and phase response curves

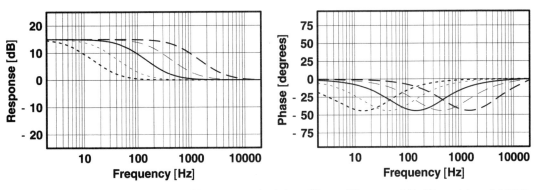

Figure 11-3. Gain and phase of low-pass shelving filter with $g = +15$ dB and $f_s = 44100$: dotted line, $f_c = 10$; thin dotted line, $f_c = 30$; solid line, $f_c = 100$; thin dashed line, $f_c = 300$; and dashed line, $f_c = 1000$

DSP Filters

using several values of cutoff frequency with a boost/cut gain g = 15 dB and f_s = 44100 Hz: dotted line, f_c = 10; thin dotted line, f_c = 30; solid line, f_c = 100; thin dashed line, f_c = 300; and dashed line, f_c = 1000.

The filter transfer function of Equation (11-1) is a precise description of the digital low-pass shelving response. Another useful set of formulas for the filter frequency response is the magnitude G(f) and the phase angle ϕ(f) of H(z):

$$G(f) = \sqrt{X^2(f) + Y^2(f)} \qquad \phi(f) = \frac{180}{\pi} \tan^{-1} \frac{Y(f)}{X(f)} \qquad (11\text{-}3a)$$

$$X(f) \equiv 1 + (\mu - 1)\cos(\phi_L(f))G_L(f) \qquad (11\text{-}3b)$$

$$Y(f) \equiv (\mu - 1)\sin(\phi_L(f))G_L(f) \qquad (11\text{-}3c)$$

where,

$$G_L(f) \equiv \sqrt{\frac{(1+\cos\theta)(1-\cos\theta'_c)}{2(1-\cos\theta\cos\theta'_c)}} \qquad \phi_L(f) \equiv -\tan^{-1}\frac{(1+\cos\theta'_c)\sin\theta}{(1+\cos\theta)\sin\theta'_c}$$

$$\theta \equiv 2\pi f / f_s \qquad \theta_c \equiv 2\pi f_c / f_s \qquad \theta'_c = 2\tan^{-1}\left[\frac{4}{1+\mu}\tan\left(\frac{\theta_c}{2}\right)\right]$$

Magnitude and phase response of the first-order low-pass shelving filter

Difference Equation

In order to implement the digital IIR filter, a difference equation is used. As previously mentioned, the IIR filter implementation is simply a running average of the input and output data:

$$u(n) = \alpha[x(n) + x(n-1)] + \gamma u(n-1) \quad (11\text{-}4a)$$

$$y(n) = x(n) + (\mu - 1)u(n) \quad (11\text{-}4b)$$

where x(n) is the current input sample x(n-1) is the previous input; u(n-1) is the previous output of the low-pass filter section; and y(n) is the

Shelving Filter

current filter output. The filter coefficients α and γ are defined in Equation (11-2), as well as in (11-5c).

```
//
// Implementation of a simple LP_Shelving Filter Stage,
// class CLP_ShelvingFilterStage

#include "cook.h"

CLP_ShelvingFilterStage::CLP_ShelvingFilterStage()
{
        x1 = 0;
        u1 = 0;
}

CLP_ShelvingFilterStage::~CLP_ShelvingFilterStage()
{
}


void CLP_ShelvingFilterStage::execute_filter_stage()
{
        double u;

        u = alpha * (x + x1) + gamma * u1;
        x1 = x;
        u1 = u;
        y = u * (mu - 1.0) + x;
}
```

$$u(n) = \alpha\left[x(n) + x(n-1)\right] + \gamma\, u(n-1) \qquad (11\text{-}5a)$$

$$y(n) = x(n) + (\mu - 1)\, u(n) \qquad (11\text{-}5b)$$

$$\gamma = \frac{1 - \left(\dfrac{4}{1+\mu}\right)\tan\dfrac{\theta_c}{2}}{1 + \left(\dfrac{4}{1+\mu}\right)\tan\dfrac{\theta_c}{2}} \qquad \alpha = (1-\gamma)/2 \qquad (11\text{-}5c)$$

$$\text{where,}\ \theta \equiv 2\pi f / f_s \qquad \theta_c \equiv 2\pi f_c / f_s$$

High-Pass IIR

Digital Filter Network

A first-order IIR high-pass shelving filter can be implemented by summing the input x(n) with the output of a first-order high-pass filter, scaled by μ-1,

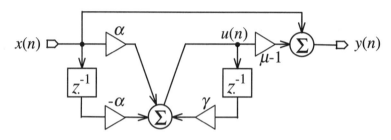

Figure 11-4. IIR high-pass shelving filter

as shown in *Figure 11-4*. The high-pass output scale factor is chosen so that when $\mu = 1$, the output is equal to the input, y(n) = x(n). The coefficient depends on the shelving level g, as $\mu \equiv 10^{g/20}$, where g is the boost/cut gain in dB.

Analogous to the low-pass shelving filter of the previous section, the high-pass shelving filter has two summing junctions, as can be seen by examining *Figure 11-4*. The left summing junction belongs to the high-pass filter subsection, which is described in detail in Chapter 5. The right summing junction forms a first-order FIR filter subsection by combining the input x(n) with the weighted value of u(n). The total filter network is first order since only one delay state is used for either the input or output states.

According to the network diagram of *Figure 11-4*, the filter transfer function for the total network can be expressed as:

$$H(z) = 1 + (\mu - 1)\frac{\alpha(1 - z^{-1})}{1 - \gamma z^{-1}} \quad (11\text{-}6)$$

Shelving Filter

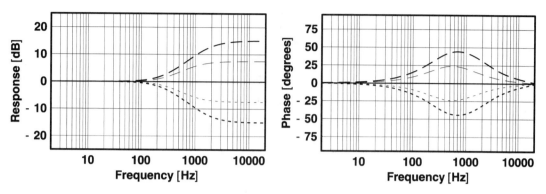

Figure 11-5. Gain and phase of high-pass shelving filter with f_c=1000 and f_s=44100 Hz: dotted line, g = -15 dB; thin dotted line, g = -7.5 dB; solid line, g = 0 dB; thin dashed line, g = +7.5 dB; and dashed line, g = +15 dB

The coefficients α and γ in Equation (11-6) are directly related to the high-pass filter center frequency f_c and the shelving filter boost/cut gain g by:

$$\gamma = \frac{1-\left(\frac{1+\mu}{4}\right)\tan\frac{\theta_c}{2}}{1+\left(\frac{1+\mu}{4}\right)\tan\frac{\theta_c}{2}} \quad (11\text{-}7)$$

$$\alpha = (1+\gamma)/2$$

where $\theta_c = 2\pi f_c / f_s$, which is the normalized cutoff frequency; and $\mu = 10^{g/20}$, where again, g is the boost/cut gain in dB.

Figure 11-5 shows several examples of gain and phase response curves for the first-order high-pass shelving filter using several values of boost/cut gain g with f_c = 1000 and f_s = 44100 Hz: dotted line, g = −15 dB; thin dotted line, g = −7.5 dB; solid line, g = 0 dB; thin dashed line, g = +7.5 dB; and dashed line, g = +15 dB. As in the previous low-pass case, the network is first order so that the phase excursion will not exceed ±90°. Again, the maximum and minimum phase value is controlled by the shelving filter boost/cut gain factor g. With the specific choice of boost/cut gains used in Figure 11-5, the phase does not exceed approximately ±45° at g = ±15 dB. Figure 11-6 shows several examples

121

DSP Filters

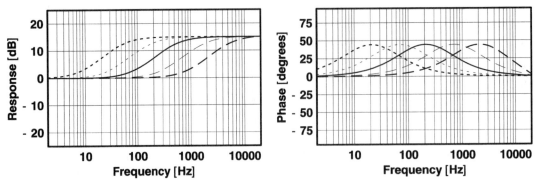

Figure 11-6. Gain and phase of high-pass shelving filter with g = +15 dB and f_s = 44100: dotted line, f_c = 30; thin dotted line, f_c = 100; solid line, f_c = 300; thin dashed line, f_c = 1000; and dashed line, f_c = 3000

of gain and phase response curves, using several values of cutoff frequency with a boost/cut gain g=15 dB and f_s = 44100 Hz: dotted line, f_c = 30; thin dotted line, f_c = 100; solid line, f_c = 300; thin dashed line, f_c = 1000; and dashed line, f_c = 3000.

The magnitude or gain response G(f) and phase response φ(f), corresponding to H(z) of the high-pass shelving filter, is described by Equations (11-8):

$$G(f) = \sqrt{X^2(f) + Y^2(f)} \qquad \phi(f) = \frac{180}{\pi} \tan^{-1} \frac{Y(f)}{X(f)} \qquad (11\text{-}8a)$$

$$X(f) \equiv 1 + (\mu - 1)\cos(\phi_H(f))G_H(f) \qquad (11\text{-}8b)$$

$$Y(f) \equiv (\mu - 1)\sin(\phi_H(f))G_H(f) \qquad (11\text{-}8c)$$

where,

$$G_H(f) \equiv \sqrt{\frac{(1 - \cos\theta)(1 + \cos\theta'_c)}{2(1 - \cos\theta\cos\theta'_c)}} \qquad \phi_H(f) \equiv \tan^{-1}\frac{(1 - \cos\theta'_c)\sin\theta}{(1 - \cos\theta)\sin\theta'_c}$$

$$\theta \equiv 2\pi f / f_s \qquad \theta_c \equiv 2\pi f_c / f_s \qquad \theta'_c = 2\tan^{-1}\left[\frac{1+\mu}{4}\tan\left(\frac{\theta_c}{2}\right)\right]$$

Magnitude and phase response of the first-order high-pass shelving filter

Shelving Filter

Difference Equation

Equations (11-9) describe the difference equation set used to implement the high-pass shelving filter:

$$u(n) = \alpha\left[x(n) - x(n-1)\right] + \gamma\, u(n-1) \quad (11\text{-}9a)$$

$$y(n) = x(n) + (\mu - 1)\, u(n) \quad (11\text{-}9b)$$

where $x(n)$ is the current input sample; $x(n\text{-}1)$ is the previous input; $u(n\text{-}1)$ is the previous output of the high-pass filter section; and $y(n)$ is the current filter output. The filter coefficients α and γ are defined in Equation (11-7), as well as in (11-10c).

$$u(n) = \alpha\left[x(n) - x(n-1)\right] + \gamma\, u(n-1) \quad (11\text{-}10a)$$

$$y(n) = x(n) + (\mu - 1)\, u(n) \quad (11\text{-}10b)$$

$$\gamma = \frac{1 - \left(\dfrac{1+\mu}{4}\right)\tan\dfrac{\theta_c}{2}}{1 + \left(\dfrac{1+\mu}{4}\right)\tan\dfrac{\theta_c}{2}} \qquad \alpha = (1+\gamma)/2 \quad (11\text{-}10c)$$

where, $\theta \equiv 2\pi f / f_s \qquad \theta_c \equiv 2\pi f_c / f_s$

Difference equation of the digital first-order high-pass shelving filter with coefficient formulas

DSP Filters

```
//
// Implementation of a simple HP_Shelving Filter Stage,
// class CHP_ShelvingFilterStage

#include "cook.h"

CHP_ShelvingFilterStage::CHP_ShelvingFilterStage()
{
      x1 = 0;
      u1 = 0;
}

CHP_ShelvingFilterStage::~CHP_ShelvingFilterStage()
{
}

void CHP_ShelvingFilterStage::execute_filter_stage()
{
      double u;

      u = alpha * (x - x1) + gamma * u1;
      x1 = x;
      u1 = u;
      y = u * (mu - 1.0) + x;
}
```

12

Cascaded Low-Pass Filter

Analog Filter Network

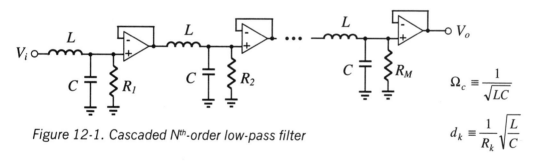

Figure 12-1. Cascaded N^{th}-order low-pass filter

$$\Omega_c \equiv \frac{1}{\sqrt{LC}}$$

$$d_k \equiv \frac{1}{R_k}\sqrt{\frac{L}{C}}$$

An Nth-order low-pass filter can be implemented by cascading the RCL circuit from Chapter 6, shown in *Figure 6-1*, where the cascaded series network makes use of the second order low-pass design formulas. In this specific implementation of the Nth order low pass, it is important that each second-order section be isolated from the remaining part of

DSP Filters

the circuit. As shown in *Figure 12-1*, op amps in the *voltage follower* configuration do a good job at isolating each second-order RCL section.

The particular form of the network in *Figure 12-1* is intended to best illustrate the connection with the digital network equivalent to be discussed in the next section. A *maximally flat* low-pass design (also known as the *Butterworth* design) determines the kth damping factor d_k. As can be seen by examining the formulas for cutoff frequency and damping factor, the product of *LC* is kept constant between sections. Therefore, in this particular design approach, only the resistor values change between sections.

The maximally flat low-pass design, or Butterworth design, has the following characteristics.
- The cutoff frequencies of each section are equal.
- The damping factors vary in a very specific way as described by Equation (12-2).

In practice, the circuit of *Figure 12-1* is not a practical way to implement an Nth-order low-pass filter. By making use of the special properties of the operational amplifier, the inductor *L* can be eliminated from each second-order section and replaced by an additional capacitor and additional resistors. The voltage follower is also replaced by another op amp configuration. Techniques of *active* filter design have been discussed at great length since the introduction of the operational amplifier in books published over the last several decades. For the purpose of this book, the RCL network of *Figure 12-1* will meet our needs.

As described in Chapter 6, the second-order transfer function for each of the sections in *Figure 12-1* is:

$$H_k(s) = \frac{\Omega_c^2}{s^2 + d_k \Omega_c s + \Omega_c^2} \quad (12\text{-}1)$$

where,

$$\Omega_c \equiv \frac{1}{\sqrt{LC}} \qquad d_k \equiv \frac{1}{R_k}\sqrt{\frac{L}{C}} \equiv 2\sin\left(\frac{(2k-1)\pi}{4M}\right) \quad (12\text{-}2)$$

Cascaded Low-Pass Filter

The particular values of R_k in Figure 12-1 are then:

$$R_k \equiv \frac{\sqrt{L/C}}{2\sin\left(\frac{(2k-1)\pi}{4M}\right)} \qquad (12\text{-}3)$$

Note the filter order $N = 2M$, where M is the number of second-order sections. The total transfer function $H(s)$ is the product of the M cascaded sections:

$$H(s) = H_1(s)\, H_2(s)\, \cdots\, H_M(s)$$

$$= \prod_{k=1}^{M} \frac{\Omega_c^2}{s^2 + d_k\, \Omega_c s + \Omega_c^2} \qquad (12\text{-}3)$$

The magnitude $G(f)$ and the phase angle $\phi(f)$ of $H(s)$ can be computed directly from $H(s)$:

$$G(f) = |H(j2\pi f)|$$

$$\phi(f) = \frac{180}{\pi} \tan^{-1}\left(\frac{\operatorname{Im}\{H(j2\pi f)\}}{\operatorname{Re}\{H(j2\pi f)\}}\right) \qquad (12\text{-}4)$$

Figure 12-2 shows several examples of gain and phase response curves for the fourth-order ($M = 2$) low-pass Butterworth filter: solid line,

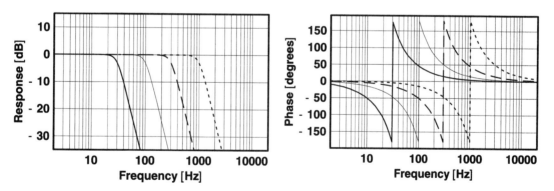

Figure 12-2. Analog gain response and phase plots for fourth-order (M=2) low-pass Butterworth filter: solid line, $f_c = 30$; thin solid line, $f_c = 100$; dashed line, $f_c = 300$; dotted line, $f_c = 1000$

DSP Filters

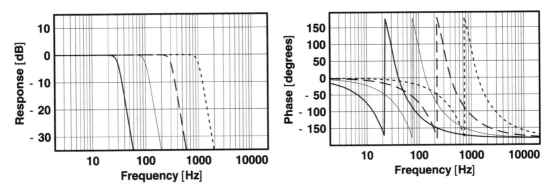

Figure 12-3. Analog gain response and phase plots for sixth-order (M=3) low-pass Butterworth filter: solid line, f_c = 30; thin solid line, f_c = 100; dashed line, f_c = 300; and dotted line, f_c = 1000

f_c = 30; thin solid line, f_c = 100; dashed line, f_c = 300; and dotted line, f_c = 1000. *Figures 12-3* and *12-4* are the sixth-order (M = 3) and eighth (M = 4) order cases, respectively. Note that the steepness of the magnitude response increases with the order of the filter, and when $f >> f_c$:

$$G(f) \to \left(\frac{1}{f^2}\right)^M = \frac{1}{f^N} \qquad (12\text{-}5)$$

Equation (12-5) corresponds to the familiar N^{th}-order low-pass filter stopband response approximation of –6 N dB/octave or –20 N dB/decade.

The phase when $f >> f_c$ goes as:

$$\varphi(f) \to -M\pi = -N \cdot 90° \qquad (12\text{-}6)$$

Because by convention a phase angle is defined to be in the range of –180° to +180°, Equation (12-6) should be modified to reflect this:

$$\varphi(f) \to \begin{cases} 0 & \text{for } M \text{ even} \\ -\pi & \text{for } M \text{ odd} \end{cases} \quad \text{when } f >> f_c \qquad (12\text{-}7)$$

Cascaded Low-Pass Filter

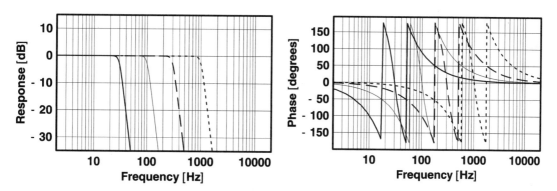

Figure 12-4. Analog gain response and phase plots for eighth-order (M=4) low-pass Butterworth filter: solid line, f_c = 30; thin solid line, f_c = 100; dashed line, f_c = 300; dotted line, f_c = 1000

The magnitude and phase response of the cascaded filter can be expressed in a way similar to the gain and phase of the second-order filter of Chapter 6. The final magnitude is the product of the gain formulas of each individual section, as shown in Equation (12-8a). Likewise, the phase is the sum of the phase angles from each section, but modified to stay within the range of −180° to 180° using the **MOD** function, as shown in Equation (12-8b).

$$G(f) = \prod_{k=1}^{M} \frac{f_c^2}{\sqrt{(f_c^2 - f^2)^2 + d_k^2 f_c^2 f^2}} \quad (12\text{-}8a)$$

$$\phi(f) = \frac{180}{\pi}\left\{\pi - \left[\left(\frac{M+2}{2}\right)\pi - \sum_{k=1}^{M} \tan^{-1}\frac{f_c^2 - f^2}{d_k f_c f}\right)\text{MOD } 2\pi\right]\right\} \quad (12\text{-}8b)$$

where,

$$f_c \equiv \frac{1}{2\pi\sqrt{LC}} \qquad d_k \equiv \frac{1}{R_k}\sqrt{\frac{L}{C}} \equiv 2\sin\left(\frac{(2k-1)\pi}{4M}\right)$$

Complex frequency response of the analog Butterworth low-pass filter

DSP Filters

NOTE: MOD or *modulo* is a standard function in most mathematical and computer languages. A common example of modulo is the measurement of time by a 12-hour clock: *clock time*, $t_c = (t \text{ MOD } 12)$, where t is absolute time. As a result, t_c is always in the range of 0 to 12, even though absolute time t can be any value.

The magnitude of the gain at the cutoff frequency can easily be determined by substituting $f = f_c$ in Equation (12-8a):

$$G(f_c) = \prod_{k=1}^{M} \frac{f_c^2}{\sqrt{(f_c^2 - f_c^2)^2 + d_k^2 f_c^2 f_c^2}} = \prod_{k=1}^{M} \frac{1}{d_k}$$

$$= \prod_{k=1}^{M} \frac{1}{2\sin\left(\frac{(2k-1)\pi}{4M}\right)} = \boxed{\frac{1}{\sqrt{2}}} \qquad (12\text{-}9a)$$

Gain response of analog N^{th}-order Buterworth low-pass filter at its cutoff frequency

Similarly, the phase response of the Butterworth filter at the cutoff frequency can be determined by substituting $f = f_c$ in Equation (12-8b):

$$\phi(f_c) = \frac{180}{\pi}\left\{\pi - \left[\left(\frac{M+2}{2}\pi - \sum_{k=1}^{M}\tan^{-1}\left(\frac{f_c^2 - f_c^2}{d_k f_c f_c}\right)\right)\text{MOD } 2\pi\right]\right\}$$

$$= \frac{180}{\pi}\left\{\pi - \left[\left(\frac{M+2}{2}\pi\right)\text{MOD } 2\pi\right]\right\}$$

$$= \begin{cases} -90° & M = 1, 5, 9, \dots \\ 180° & M = 2, 6, 10, \dots \\ 90° & M = 3, 7, 11, \dots \\ 0° & M = 4, 8, 12, \dots \end{cases} \qquad (12\text{-}9b)$$

Phase response of analog N^{th}-order Butterworth low-pass filter at its cutoff frequency

Cascaded Low-Pass Filter

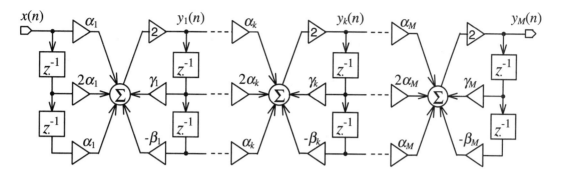

Figure 12-5. Cascaded N^{th}-order digital low-pass filter network

Digital Filter Network

As shown in *Figure 12-5*, a digital N^{th}-order low-pass filter can be implemented by cascading the second-order network from Chapter 6, *Figure 6-4*. Each individual section of the cascaded series makes use of the second-order low-pass design formulas presented previously in Chapter 6. A *maximally flat* low-pass design (Butterworth design) determines the kth damping factor d_k. As discussed in the previous section, the maximally flat, Butterworth design, has:

- equal cutoff frequencies in each section.
- the damping factors vary in a very specific way, described by Equation (12-11).

As discussed in Chapter 6, the second-order transfer function for each of the sections in Figure 12-5 is:

$$H_k(z) = \frac{\alpha_k \left(1 + 2z^{-1} + z^{-2}\right)}{\frac{1}{2} - \gamma_k z^{-1} + \beta_k z^{-2}} \quad (12\text{-}9)$$

The coefficients α_k, β_k, and γ_k in Equation (12-9) are directly related to the filter damping factor d_k and cutoff frequency f_c by:

$$\beta_k = \frac{1}{2}\left(\frac{1-\tfrac{1}{2}d_k \sin\theta_c}{1+\tfrac{1}{2}d_k \sin\theta_c}\right)$$

$$\gamma_k = \left(\tfrac{1}{2}+\beta_k\right)\cos\theta_c \qquad (12\text{-}10)$$

where,
$$\alpha_k = \left(\tfrac{1}{2}+\beta_k-\gamma_k\right)/4$$

$$d_k \equiv 2\sin\left(\frac{(2k-1)\pi}{4M}\right) \qquad (12\text{-}11)$$

$$\theta_c \equiv 2\pi f_c / f_s$$

Note the filter order $N = 2M$, where M is the number of second-order sections. The total transfer function $H(z)$ is the product of the M cascaded sections:

$$H(z)H_1(z)H_2(z)\cdots H_M(z)$$
$$= \prod_{k=1}^{M} \frac{\alpha_k\left(1+2z^{-1}+z^{-2}\right)}{\tfrac{1}{2}-\gamma_k z^{-1}+\beta_k z^{-2}} \qquad (12\text{-}12)$$

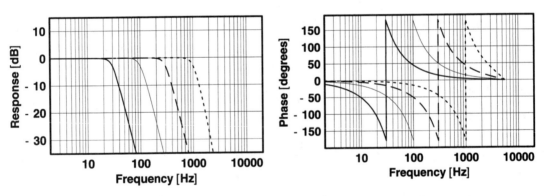

Figure 12-6. Gain and phase response of digital fourth-order (M=2) low-pass Butterworth filter with $f_s = 11025$: solid line, $f_c = 30$; thin solid line, $f_c = 100$; dashed line, $f_c = 300$; dotted line, $f_c = 1000$

Cascaded Low-Pass Filter

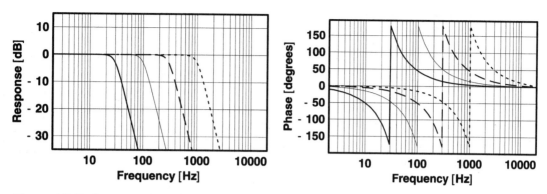

Figure 12-7. Gain and phase response of digital fourth-order (M=2) low-pass Butterworth filter with f_s = 44100: solid line, f_c = 30; thin solid line, f_c = 100; dashed line, f_c = 300; dotted line, f_c = 1000

The magnitude G(f) and the phase angle φ(f) of H(z) can be computed directly from H(z):

$$G(f) = |H(e^{j\theta})| \qquad \phi(f) = \frac{180}{\pi} \tan^{-1}\left(\frac{\text{Im}\{H(e^{j\theta})\}}{\text{Re}\{H(e^{j\theta})\}}\right) \qquad (12\text{-}13)$$

where, $\theta \equiv 2\pi f / f_s$ is the *normalized frequency*.

Figure 12-6 shows several examples of gain and phase response curves for the fourth-order (M = 2) low-pass Butterworth filter: solid line, f_c = 30; thin solid line, f_c = 100; dashed line, f_c = 300; and dotted line, f_c = 1000. In this case, the sample frequency f_s = 11025 Hz, while the Nyquist is half of that, or f_n = 5512.5 Hz. *Figures 12-7* are identical filter conditions as those in the previous figure, but the sample frequency has been increased by a factor of four, so that f_s = 44100 Hz and f_n = 22050 Hz. *Figures 12-8* are again identical filter conditions as those in the Figure 12-6, but the filter order is N = 6 (M = 3).

Note that as with the analog filter, the steepness of the magnitude response increases with the order of the filter, and when $f_c \ll f \ll f_n$:

$$G(f) \to \begin{cases} \dfrac{1}{f^N} & f_c \ll f \ll f_n \\ 0 & f \to f_n \end{cases} \qquad (12\text{-}14)$$

DSP Filters

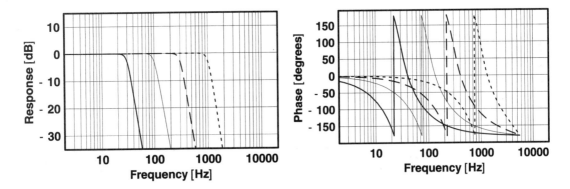

Figure 12-8. Gain and phase response of digital sixth-order (M=3) low-pass Butterworth filter with f_s = 11025: solid line, f_c = 30; thin solid line, f_c = 100; dashed line, f_c = 300; and dotted line, f_c = 1000

Equation (12-14) corresponds to the familiar N^{th}-order low-pass filter stopband response approximation of $-6N$ dB/octave or $-20N$ dB/decade when f is much greater than the cutoff frequency, as well as much less than the Nyquist frequency. As f approaches the Nyquist frequency, the filter roll-off steepness increases and the gain becomes exactly zero when $f = f_n$. In some ways, this characteristic of the digital low-pass filter (also true with the first- and second-order filters) gives an extra boost of stopband suppression as compared to the equivalent analog filters.

The phase when $f_c \ll f \leq f_n$ goes as:

$$\varphi(f) \to -M\pi = -N \cdot 90° \quad (12\text{-}15)$$

Because by convention, as in the analog case, a phase angle is defined to be in the range of $-180°$ to $+180°$, Equation (12-15) should be modified to reflect this:

$$\varphi(f) \to \begin{cases} 0 & \text{for } M \text{ even} \\ -\pi & \text{for } M \text{ odd} \end{cases} \quad \text{when } f_c \ll f \leq f_n \quad (12\text{-}16)$$

Cascaded Low-Pass Filter

The magnitude and phase response of the cascaded filter can be expressed in a way similar to the gain and phase of the second-order digital filter of Chapter 6. The final magnitude is the product of the gain formulas of each individual section as shown in Equation (12-17a). Likewise, the phase is the sum of the phase angles from each section, but modified to stay within the range of −180° to 180° using the **MOD** function as shown in Equation (12-17b).

$$G(f) = \prod_{k=1}^{M} \frac{(1+\cos\theta)(1-\cos\theta_c)}{\sqrt{(d_k \sin\theta \sin\theta_c)^2 + 4(\cos\theta - \cos\theta_c)^2}} \quad (12\text{-}17a)$$

$$\phi(f) = \frac{180}{\pi}\left\{\pi - \left[\left(\frac{M+2}{2}\right)\pi - \sum_{k=1}^{M}\tan^{-1}\left(\frac{2(\cos\theta - \cos\theta_c)}{d_k \sin\theta \sin\theta_c}\right)\right] \text{MOD } 2\pi\right\} \quad (12\text{-}17b)$$

where,

$$d_k \equiv 2\sin\left(\frac{(2k-1)\pi}{4M}\right) \quad (12\text{-}17c)$$

$$\theta_c \equiv 2\pi f_c / f_s \qquad \theta \equiv 2\pi f / f_s$$

Magnitude and phase response formulas for the Nth-order IIR low-pass Butterworth filter

The magnitude of the gain at the cutoff frequency can easily be determined by substituting $f = f_c$ in Equation (12-17a):

$$G(f_c) = \prod_{k=1}^{M} \frac{(1+\cos\theta_c)(1-\cos\theta_c)}{\sqrt{(d_k \sin\theta_c \sin\theta_c)^2 + 4(\cos\theta_c - \cos\theta_c)^2}}$$

$$= \prod_{k=1}^{M} \frac{1}{d_k} = \prod_{k=1}^{M} \frac{1}{2\sin\left(\frac{(2k-1)\pi}{4M}\right)} = \frac{1}{\sqrt{2}} \quad (12\text{-}18a)$$

Gain response of Nth-order IIR Butterworth low-pass filter at its cutoff frequency

DSP Filters

Similarly, the phase response of the Butterworth filter at the cutoff frequency can be determined by substituting $f = f_c$ in Equation (12-17b):

$$\phi(f_c) = \frac{180}{\pi}\left\{\pi - \left[\left(\frac{M+2}{2}\pi - \sum_{k=1}^{M}\tan^{-1}\frac{2(\cos\theta_c - \cos\theta_c)}{d_k \sin\theta_c \sin\theta_c}\right)\text{MOD } 2\pi\right]\right\}$$

$$= \frac{180}{\pi}\left\{\pi - \left[\left(\frac{M+2}{2}\pi\right)\text{MOD } 2\pi\right]\right\}$$

$$= \begin{cases} -90° & M = 1, 5, 9, \ldots \\ 180° & M = 2, 6, 10, \ldots \\ 90° & M = 3, 7, 11, \ldots \\ 0° & M = 4, 8, 12, \ldots \end{cases} \qquad (12\text{-}18\text{b})$$

Phase response of Nth-order IIR Butterworth low-pass filter at its cutoff frequency

Difference Equation

In order to implement the Nth-order IIR filter, a set of *difference equations* are needed. As previously discussed, the IIR filter implementation is simply a running average of the input and output data for each of the M second-order sections, where M difference equations are used, corresponding to each of the M summing junctions:

$$y_k(n) = 2\{\alpha_k[x_k(n) + 2x_k(n-1) + x_k(n-2)] + \gamma_k \, y_k(n-1) - \beta_k \, y_k(n-2)\} \quad (12\text{-}18)$$

where the input is $x_1(n) = x(n)$ and the output is $y(n) = y_M(n)$.

Cascaded Low-Pass Filter

$$y_k(n) = 2\{\alpha_k [x_k(n) + 2x_k(n-1) + x_k(n-2)] + \gamma_k\, y_k(n-1) - \beta_k\, y_k(n-2)\} \quad (12\text{-}19a)$$

$$\beta_k = \frac{1}{2}\left(\frac{1 - \tfrac{1}{2} d_k \sin\theta_c}{1 + \tfrac{1}{2} d_k \sin\theta_c}\right) \qquad \gamma_k = \left(\tfrac{1}{2} + \beta_k\right)\cos\theta_c \qquad \alpha_k = \left(\tfrac{1}{2} + \beta_k - \gamma_k\right)/4 \quad (12\text{-}19b)$$

$$d_k = 2\sin\left(\frac{(2k-1)\pi}{4M}\right) \quad (12\text{-}19c)$$

where, $\theta \equiv 2\pi f / f_s \qquad \theta_c \equiv 2\pi f_c / f_s$

Difference equations of Nth-order IIR low-pass Butterworth filter with coefficient formulas

Partial C++ code to implement Equation (12-18) follows:

```cpp
//
// Implementation of Cascaded Lowpass Filter Block,
// class CCascaded_Lowpass
 void CCascaded_Lowpass::execute_filter_block_in_place(double *in)
{
        int i;
        double input = *in;

        stages[0]->x = input;
        stages[0]->execute_filter_stage();

        for (i= 1; i< NUM_BANDS; i++)
        {
                stages[i]->x = stages[i-1]->y;
                stages[i]->execute_filter_stage();
        }
        *in = stages[NUM_BANDS-1]->y;
}

//
// cook_lowpass.cpp
//
// Implementation of a simple Lowpass Filter Stage,
// class CLowpassFilterStage
```

```
#include "cook.h"

CLowpassFilterStage::CLowpassFilterStage()
{
    x1 = 0;
    x2 = 0;
    y1 = 0;
    y2 = 0;
}

CLowpassFilterStage::~CLowpassFilterStage()
{
}

void CLowpassFilterStage::execute_filter_stage()
{
    y = 2 * (alpha * (x + 2.0 * x1+ x2) + gamma * y1 - beta * y2);
    x2 = x1;
    x1 = x;
    y2 = y1;
    y1 = y;
}
```

13

Cascaded High-Pass Filter

Analog Filter Network

As with the low-pass filter of the previous chapter, an Nth-order high-pass filter can be implemented by cascading the RCL circuit of *Figure 7-1*, making use of the second-order high-pass design formulas. The particular form of the network in *Figure 13-1* is intended to best illustrate the connection with the digital network equivalent, even though in practice, the circuit of *Figure 13-1* is not a practical way to implement an Nth-order

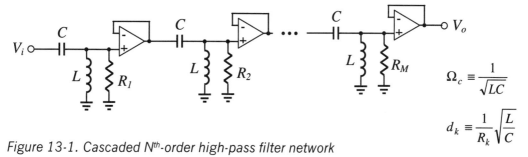

Figure 13-1. Cascaded N^{th}-order high-pass filter network

$$\Omega_c \equiv \frac{1}{\sqrt{LC}}$$

$$d_k \equiv \frac{1}{R_k}\sqrt{\frac{L}{C}}$$

DSP Filters

high-pass filter. (By making use of the special properties of the operational amplifier, the inductor L can be eliminated from each second-order section and replaced by an additional capacitor and additional resistors along with a different op amp configuration).

The Butterworth (maximally flat) filter design has the following characteristics, as described by Equation (13-2):
- The product of LC is kept constant between sections.
- Only the resistor values change between sections.
- The cutoff frequencies of each section are equal.
- The damping factors d_k of each kth section depend specifically on the order N and the section number k.

As described in Chapter 7, the second order transfer function for each of the sections in *Figure 13-1* is:

$$H_k(s) = \frac{s^2}{s^2 + d_k \Omega_c s + \Omega_c^2} \quad (13\text{-}1)$$

where,

$$\Omega_c \equiv \frac{1}{\sqrt{LC}} \qquad d_k \equiv \frac{1}{R_k}\sqrt{\frac{L}{C}} \equiv 2\sin\left(\frac{(2k-1)\pi}{4M}\right) \quad (13\text{-}2)$$

The particular values of R_k in *Figure 13-1* are then:

$$R_k \equiv \frac{\sqrt{L/C}}{2\sin\left(\frac{(2k-1)\pi}{4M}\right)} \quad (13\text{-}3)$$

Note the filter order N = 2M, where M is the number of second-order sections. The total transfer function H(s) is the product of the M cascaded sections:

$$H(s) = H_1(s)\ H_2(s)\ \cdots\ H_M(s)$$

$$= \prod_{k=1}^{M} \frac{s^2}{s^2 + d_k \Omega_c s + \Omega_c^2} \quad (13\text{-}3)$$

Cascaded High-Pass Filter

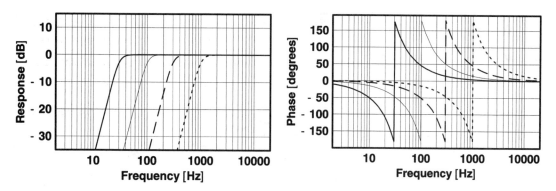

Figure 13-2. Analog gain response and phase plots for fourth-order (M=2) high-pass Butterworth filter: solid line, f_c = 30; thin solid line, f_c = 100; dashed line, f_c = 300; dotted line, f_c = 1000

The magnitude $G(f)$ and the phase angle $\phi(f)$ of $H(s)$ can be computed directly from $H(s)$:

$$G(f) = |H(j2\pi f)|$$

$$\phi(f) = \frac{180}{\pi} \tan^{-1}\left(\frac{\text{Im}\{H(j2\pi f)\}}{\text{Re}\{H(j2\pi f)\}}\right) \quad (13\text{-}4)$$

Figure 13-2 shows several examples of gain and phase response curves for the fourth-order ($M = 2$) high-pass Butterworth filter: solid line, f_c = 30; thin solid line, f_c = 100; dashed line, f_c = 300; and dotted line, f_c = 1000. *Figures 13-3* and *13-4* are the sixth-order ($M = 3$) and eighth ($M = 4$) order cases, respectively. Note that the steepness of the magnitude response increases with the order of the filter and when $f << f_c$:

$$G(f) \to \left(f^2\right)^M = f^N \quad (13\text{-}5)$$

Equation (13-5) corresponds to the familiar N^{th}-order high-pass filter stopband response approximation of $+6\,N$ dB/octave or $+20\,N$ dB/decade.

DSP Filters

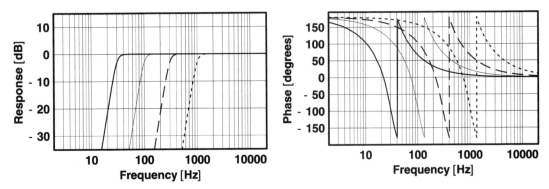

Figure 13-3. Analog gain response and phase plots for sixth-order (M=3) high-pass Butterworth filter: solid line, $f_c = 30$; thin solid line, $f_c = 100$; dashed line, $f_c = 300$; and dotted line, $f_c = 1000$

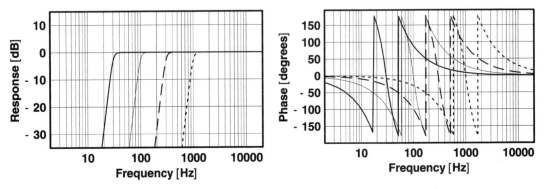

Figure 13-4. Analog gain response and phase plots for eighth-order (M=4) high-pass Butterworth filter: solid line, $f_c = 30$; thin solid line, $f_c = 100$; dashed line, $f_c = 300$; and dotted line, $f_c = 1000$.

The phase when $f \ll f_c$ goes as:

$$\varphi(f) \to M\pi = N \cdot 90° \qquad (13\text{-}6)$$

Using the convention that a phase angle falls in the range of $-180°$ to $+180°$, Equation (13-6) can be modified to reflect this definition:

$$\varphi(f) \to \begin{cases} 0 & \text{for } M \text{ even} \\ \pi & \text{for } M \text{ odd} \end{cases} \quad \text{when } f \ll f_c \qquad (13\text{-}7)$$

Cascaded High-Pass Filter

The magnitude and phase response of the cascaded filter can be expressed in a way similar to the gain and phase of the second-order filter of Chapter 7. The final magnitude is the product of the gain formulas of each individual section as shown in Equation (13-8a). Similarly, the phase is the sum of the phase angles from each section modified to stay within the range of −180° to 180° using the **MOD** function as shown in Equation (13-8b).

$$G(f) = \prod_{k=1}^{M} \frac{f^2}{\sqrt{(f_c^2 - f^2)^2 + d_k^2 f_c^2 f^2}} \quad (13\text{-}8a)$$

$$\phi(f) = \frac{180}{\pi} \left\{ \pi - \left[\left(\frac{-M+2}{2} \pi - \sum_{k=1}^{M} \tan^{-1} \frac{f_c^2 - f^2}{d_k f_c f} \right) \text{MOD } 2\pi \right] \right\} \quad (13\text{-}8b)$$

where,
$$f_c \equiv \frac{1}{2\pi\sqrt{LC}} \qquad d_k \equiv \frac{1}{R_k}\sqrt{\frac{L}{C}} \equiv 2\sin\left(\frac{(2k-1)\pi}{4M}\right)$$

Complex frequency response of the analog Butterworth high-pass filter

The magnitude of the gain at the cutoff frequency can easily be determined by substituting $f = f_c$ in Equation (13-8a):

DSP Filters

$$G(f_c) = \prod_{k=1}^{M} \frac{f_c^2}{\sqrt{(f_c^2 - f_c^2)^2 + d_k^2 f_c^2 f_c^2}} = \prod_{k=1}^{M} \frac{1}{d_k}$$

$$= \prod_{k=1}^{M} \frac{1}{2\sin\left(\frac{(2k-1)\pi}{4M}\right)} = \boxed{\frac{1}{\sqrt{2}}} \qquad (13\text{-}9a)$$

Gain response of the analog N^{th}-order Butterworth high-pass filter at its cutoff frequency

Similarly, the phase response of the Butterworth filter at the cutoff frequency can be determined by substituting $f = f_c$ in Equation (13-8b):

$$\phi(f_c) = \frac{180}{\pi}\left\{\pi - \left[\left(\frac{-M+2}{2}\pi - \sum_{k=1}^{M}\tan^{-1}\frac{f_c^2 - f_c^2}{d_k f_c f_c}\right) \text{MOD } 2\pi\right]\right\}$$

$$= \frac{180}{\pi}\left\{\pi - \left[\left(\frac{-M+2}{2}\pi\right) \text{MOD } 2\pi\right]\right\}$$

$$= \boxed{\begin{cases} 90° & M = 1, 5, 9, \ldots \\ 180° & M = 2, 6, 10, \ldots \\ -90° & M = 3, 7, 11, \ldots \\ 0° & M = 4, 8, 12, \ldots \end{cases}} \qquad (13\text{-}9b)$$

Phase response of the analog N^{th}-order Butterworth high-pass filter at its cutoff frequency

Digital Filter Network

A digital N^{th}-order high-pass filter can be implemented by cascading the second-order network from *Figure 7-4*. Each individual section of the cascaded series makes use of the second-order high-pass design formulas presented in Chapter 7. Note that each section of the cascaded (*direct form*) network of *Figure 13-1* shares a set of delay elements. In this way, the total number of delay elements for the entire network is approximately equal to the order of the filter for large filter order. In contrast, the single second-order (direct form) network uses four delay elements.

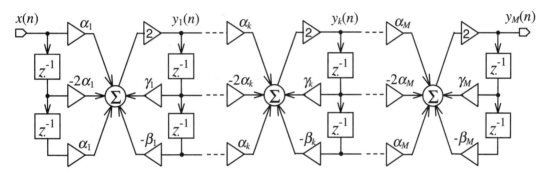

Figure 13-5. Cascaded Nth-order digital high-pass filter network

The second-order transfer function from Chapter 7, which is used for each of the sections in *Figure 13-5* is:

$$H_k(z) = \frac{\alpha_k \left(1 - 2 z^{-1} + z^{-2}\right)}{\frac{1}{2} - \gamma_k z^{-1} + \beta_k z^{-2}} \quad (13\text{-}9)$$

The coefficients α_k, β_k, and γ_k in Equation (13-9) are directly related to the filter damping factor d_k and cutoff frequency f_c by:

$$\beta_k = \frac{1}{2}\left(\frac{1-\frac{1}{2}d_k \sin\theta_c}{1+\frac{1}{2}d_k \sin\theta_c}\right)$$

$$\gamma_k = \left(\tfrac{1}{2}+\beta_k\right)\cos\theta_c \qquad (13\text{-}10)$$

$$\alpha_k = \left(\tfrac{1}{2}+\beta_k+\gamma_k\right)/4$$

where,

$$d_k \equiv 2\sin\left(\frac{(2k-1)\pi}{4M}\right) \qquad (13\text{-}11)$$

$$\theta_c \equiv 2\pi f_c / f_s$$

Note the filter order $N = 2M$ where M is the number of second-order sections. The total transfer function $H(z)$ is the product of the M cascaded sections:

$$H(z) = H_1(z)H_2(z)\cdots H_M(z) \qquad (13\text{-}12)$$
$$= \prod_{k=1}^{M} \frac{\alpha_k\left(1-2z^{-1}+z^{-2}\right)}{\tfrac{1}{2}-\gamma_k z^{-1}+\beta_k z^{-2}}$$

The magnitude $G(f)$ and the phase angle $\phi(f)$ of $H(z)$ can be computed directly from $H(z)$:

$$G(f) = |H(e^{j\theta})| \qquad \phi(f) = \frac{180}{\pi}\tan^{-1}\left(\frac{\text{Im}\{H(e^{j\theta})\}}{\text{Re}\{H(e^{j\theta})\}}\right) \qquad (13\text{-}13)$$

where, $\theta \equiv 2\pi f / f_s$ is the normalized frequency.

Figure 13-6 shows several examples of gain and phase response curves for the fourth-order ($M = 2$) high-pass Butterworth filter: solid line, $f_c = 30$; thin solid line, $f_c = 100$; dashed line, $f_c = 300$; and dotted line, $f_c = 1000$. In this case, the sample frequency $f_s = 11025$ Hz, while the

Cascaded High-Pass Filter

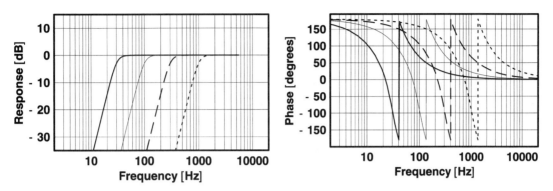

Figure 13-6. Gain and phase response of digital fourth-order (M=2) high-pass Butterworth filter with f_s = 11025: solid line, f_c = 30; thin solid line, f_c = 100; dashed line, f_c = 300; dotted line, f_c = 1000

Nyquist is half of that, or f_n = 5512.5 Hz. *Figures 13-7* and *13-8* are equivalent specifications to those in the previous figure, but the filter order is N = 6 (M = 3) and N = 8 (M = 4), respectively.

Note that as with the analog filter, the steepness of the magnitude response increases with the order of the filter, and when $f \ll f_c \ll f_n$:

$$G(f) \to f^N \quad f \ll f_c \ll f_n \quad (13\text{-}14)$$

Equation (13-14) corresponds to the Nth-order high-pass filter stopband response approximation of +6N dB/octave or +20N dB/decade when f is much less than the cutoff frequency and when the cutoff frequency is much less than the Nyquist frequency.

The phase when $f \ll f_c \leq f_n$ goes as:

$$\varphi(f) \to M\pi = N \cdot 90° \quad (13\text{-}15)$$

Because by convention, as in the analog case, a phase angle is defined to be in the range of −180° to +180°, the Equation (13-6) should be modified to reflect this:

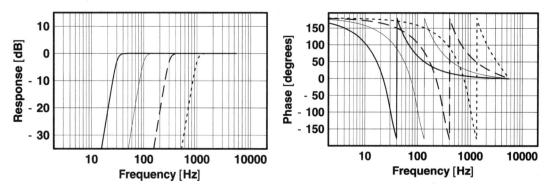

Figure 13-7. Gain and phase response of digital sixth-order (M=3) high-pass Butterworth filter with $f_s = 11025$: solid line, $f_c = 30$; thin solid line, $f_c = 100$; dashed line, $f_c = 300$; dotted line, $f_c = 1000$

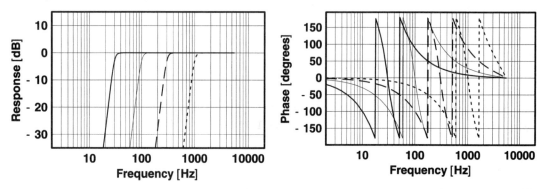

Figure 13-8. Gain and phase response of digital eighth-order (M=4) high-pass Butterworth filter with $f_s = 11025$: solid line, $f_c = 30$; thin solid line, $f_c = 100$; dashed line, $f_c = 300$; dotted line, $f_c = 1000$

$$\varphi(f) \to \begin{cases} 0 & M \text{ even} \\ \pi & M \text{ odd} \end{cases} \quad (13\text{-}16)$$

for $f \ll f_c \leq f_n$.

The magnitude and phase response of the cascaded filter can be expressed in a way similar to the gain and phase of the second-order digital filter of Chapter 7. The final magnitude is the product of the gain formulas of each individual section as shown in Equation (13-17a). Like-

Cascaded High-Pass Filter

wise, the phase is the sum of the phase angles from each section, but modified to stay within the range of −180° to 180° using the **MOD** function, as shown in Equation (13-17b).

$$G(f) = \prod_{k=1}^{M} \frac{(1-\cos\theta)(1+\cos\theta_c)}{\sqrt{(d_k \sin\theta \sin\theta_c)^2 + 4(\cos\theta - \cos\theta_c)^2}} \quad (13\text{-}17a)$$

$$\phi(f) = \frac{180}{\pi}\left\{\pi - \left[\left(\frac{-M+2}{2}\pi - \sum_{k=1}^{M} \tan^{-1}\frac{2(\cos\theta - \cos\theta_c)}{d_k \sin\theta \sin\theta_c}\right) \text{MOD } 2\pi\right]\right\} \quad (13\text{-}17b)$$

$$d_k \equiv 2\sin\left(\frac{(2k-1)\pi}{4M}\right) \quad (13\text{-}17c)$$

where,

$$\theta_c \equiv 2\pi f_c / f_s \qquad \theta \equiv 2\pi f / f_s$$

Magnitude and phase response formulas for the Nth-order IIR high-pass Butterworth filter

The magnitude of the gain at the cutoff frequency can be determined by substituting $f = f_c$ in Equation (13-17a):

$$G(f_c) = \prod_{k=1}^{M} \frac{(1-\cos\theta_c)(1+\cos\theta_c)}{\sqrt{(d_k \sin\theta_c \sin\theta_c)^2 + 4(\cos\theta_c - \cos\theta_c)^2}}$$

$$= \prod_{k=1}^{M} \frac{1}{d_k} = \prod_{k=1}^{M} \frac{1}{2\sin\left(\frac{(2k-1)\pi}{4M}\right)} = \boxed{\frac{1}{\sqrt{2}}} \quad (13\text{-}18a)$$

Gain response of Nth-order IIR Butterworth high-pass filter at its cutoff frequency

DSP Filters

Similarly, the phase response of the Butterworth filter at the cutoff frequency can be determined by substituting $f = f_c$ in Equation (13-17b):

$$\phi(f_c) = \frac{180}{\pi}\left\{\pi - \left[\left(\frac{-M+2}{2}\pi - \sum_{k=1}^{M}\tan^{-1}\frac{2(\cos\theta_c - \cos\theta_c)}{d_k \sin\theta_c \sin\theta_c}\right) \text{MOD } 2\pi\right]\right\}$$

$$= \frac{180}{\pi}\left\{\pi - \left[\left(\frac{-M+2}{2}\pi\right) \text{MOD } 2\pi\right]\right\}$$

$$= \begin{cases} 90° & M = 1, 5, 9, \ldots \\ 180° & M = 2, 6, 10, \ldots \\ -90° & M = 3, 7, 11, \ldots \\ 0° & M = 4, 8, 12, \ldots \end{cases} \qquad (13\text{-}18b)$$

Phase response of Nth-order IIR Butterworth high-pass filter at its cutoff frequency

Difference Equation

As previously discussed, the IIR filter implementation is simply a running average of the input and output data for each of the M second-order sections, where M difference equations are used, corresponding to each of the M summing junctions:

$$y_k(n) = 2\{\alpha_k [x_k(n) - 2x_k(n-1) + x_k(n-2)] + \gamma_k\, y_k(n-1) - \beta_k\, y_k(n-2)\} \quad (13\text{-}18)$$

where the input is $x_1(n) = x(n)$ and the output is $y(n) = y_M(n)$.

Cascaded High-Pass Filter

$$y_k(n) = 2\{\alpha_k [x_k(n) - 2x_k(n-1) + x_k(n-2)] + \gamma_k\, y_k(n-1) - \beta_k\, y_k(n-2)\} \quad (13\text{-}19a)$$

$$\beta_k = \frac{1}{2}\left(\frac{1 - \frac{1}{2} d_k \sin\theta_c}{1 + \frac{1}{2} d_k \sin\theta_c}\right) \quad \gamma_k = \left(\tfrac{1}{2} + \beta_k\right)\cos\theta_c \quad \alpha_k = \left(\tfrac{1}{2} + \beta_k + \gamma_k\right)/4 \quad (13\text{-}19b)$$

$$d_k = 2\sin\left(\frac{(2k-1)\pi}{4M}\right) \quad (13\text{-}19b)$$

$$\text{where,} \quad \theta \equiv 2\pi f / f_s \quad \theta_c \equiv 2\pi f_c / f_s$$

Difference equations of Nth-order IIR high-pass Butterworth filter with coefficient formulas

Partial C++ code to implement Equation (13-18) follows:

```
//
// Implementation of Cascaded Highpass Filter Block,
// class CCascaded_Highpass
void CCascaded_Highpass::execute_filter_block_in_place(double *in)
{
        int i;
        double input = *in;

        stages[0]->x = input;
        stages[0]->execute_filter_stage();

        for (i= 1; i< NUM_BANDS; i++)
        {
                stages[i]->x = stages[i-1]->y;
                stages[i]->execute_filter_stage();
        }

        *in = stages[NUM_BANDS-1]->y;
}
//
// cook_highpass.cpp
//
// Implementation of a simple Highpass Filter Stage,
// class CHighpassFilterStage

#include "cook.h"
```

DSP Filters

```
CHighpassFilterStage::CHighpassFilterStage()
{
      x1 = 0;
      x2 = 0;
      y1 = 0;
      y2 = 0;
}

CHighpassFilterStage::~CHighpassFilterStage()
{
}

void CHighpassFilterStage::execute_filter_stage()
{
      y = 2 * (alpha * (x - 2.0 * x1+ x2) + gamma * y1 - beta * y2);
      x2 = x1;
      x1 = x;
      y2 = y1;
      y1 = y;
}
```

14

CASCADED

BANDPASS FILTER

Digital Filter Network

As shown in *Figure 14-1a* and *14-1b,* a digital N^{th}-order bandpass filter can be implemented by cascading the second-order network, which was discussed in Chapter 8, where the second-order transfer function for each of the M cascaded sections is expressed as:

$$H_i(z) = \frac{\alpha_i \left(1 - z^{-2}\right)}{\frac{1}{2} - \gamma_i z^{-1} + \beta_i z^{-2}} \quad (14\text{-}1)$$

The total transfer function H(z) is the product of the M cascaded sections, where the filter order N = 2M:

$$H(z) = H_1(z) H_2(z) \cdots H_M(z) \quad (14\text{-}2a)$$

Instead of a product of M sections as described above, the network can be expressed as M/2 pairs of cascaded sections:

DSP Filters

$$H(z) = H_{11}(z)\, H_{21}(z)\, H_{12}(z)\, H_{22}(z) \cdots H_{1\frac{M}{2}}(z)\, H_{2\frac{M}{2}}(z)$$

$$= \prod_{k=1}^{M/2} \prod_{j=1}^{2} H_{jk}(z) = \prod_{k=1}^{M/2} \prod_{j=1}^{2} \frac{\alpha_{jk}(1 - z^{-2})}{\frac{1}{2} - \gamma_{jk}\, z^{-1} + \beta_{jk}\, z^{-2}} \quad (14\text{-}2b)$$

The number of product terms in Equation (14-2b) is still M, but now each pair of $j = 1$ and $j = 2$ products correspond to a $k = 1, 2, \ldots M/2$ grouping. The reason for this organization of $H(z)$ may be more obvious by examining *Figure 14-2* : $j = 1$ corresponds to a second-order bandpass filter whose center frequency is below f_0, while $j = 2$ corresponds to a bandpass whose center frequency is above f_0. Each pair of k sections has a gain value different than one, which is shared by both the $j = 1$ and $j = 2$ sections. This arrangement leads to a *maximally flat* bandpass design (*Butterworth* design).

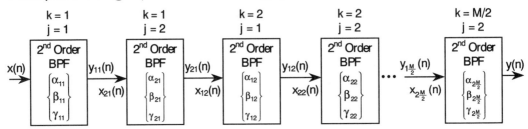

Figure 14-1a. Cascaded Nth-order IIR bandpass filter network

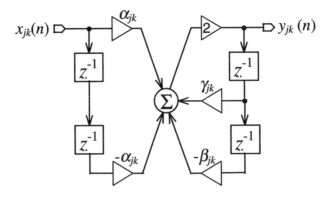

Figure 14-1b. Single second-order section of cascaded bandpass filter network

Cascaded Bandpass Filter

The coefficients α_{jk}, β_{jk}, and γ_{jk} in Equation (14-2b) are directly related to the filter Q and center frequency f_0 of the total network. However, unlike the coefficient formulas of the low-pass and high-pass cascaded Butterworth filters, the cascaded bandpass coefficient formulas are somewhat elaborate:

$$\beta_{jk} = \frac{1}{2} \frac{1 - \frac{1}{2} d_k \sin\theta_{jk}}{1 + \frac{1}{2} d_k \sin\theta_{jk}} \quad (14\text{-}3a)$$

$$\gamma_{jk} = \left(\tfrac{1}{2} + \beta_{jk}\right)\cos\theta_{jk} \quad (14\text{-}3b)$$

$$\alpha_{jk} = \tfrac{1}{2}\left(\tfrac{1}{2} - \beta_{jk}\right)\sqrt{1 + \left(\frac{W_k - 1/W_k}{d_k}\right)^2} \quad (14\text{-}3c)$$

$$d_k = \sqrt{\frac{d_E D_k}{A_k + \sqrt{A_k^2 - 1}}} \quad (14\text{-}3d)$$

$$d_E = \frac{2\tan\left(\dfrac{\theta_0}{2Q}\right)}{\sin\theta_0} \quad (14\text{-}3e)$$

$$D_k = 2\sin\left(\frac{(2k-1)\pi}{2M}\right) \quad (14\text{-}3f) \qquad A_k = \frac{1 + (d_E/2)^2}{D_k \, d_E/2} \quad (14\text{-}3g)$$

$$W_k = B_k + \sqrt{B_k^2 - 1} \quad (14\text{-}3h) \qquad B_k = \frac{D_k}{d_k}(d_E/2) \quad (14\text{-}3i)$$

$$\theta_{1k} = 2\tan^{-1}\left\{\frac{\tan(\theta_0/2)}{W_k}\right\} \quad (14\text{-}3j) \qquad \theta_{2k} = 2\tan^{-1}\{W_k \tan(\theta_0/2)\} \quad (14\text{-}3k)$$

where,

$$\theta \equiv 2\pi f/f_s \qquad \theta_0 \equiv 2\pi f_0/f_s \quad (14\text{-}3l)$$

DSP Filters

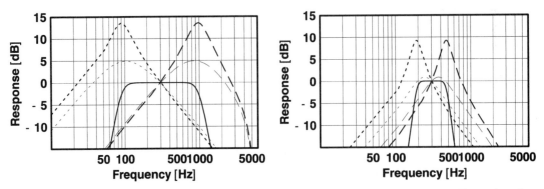

Figure 14-2. Gain response of eighth-order (M=4) Butterworth bandpass filter showing equivalent gain response of each individual second-order section: $f_0 = 300$ Hz; left, $Q = 0.33$; right, $Q = 1.0$. Solid line, total H(z); dotted line $H_{11}(z)$; dashed line $H_{21}(z)$; thin dotted line $H_{12}(z)$; thin dashed line $H_{22}(z)$

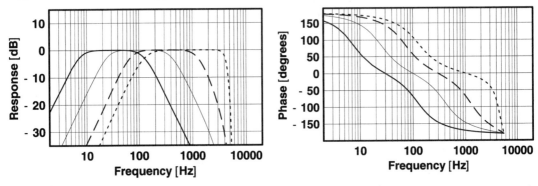

Figure 14-3. Gain and phase response of digital fourth-order (M=2) bandpass Butterworth filter with $f_s = 11025$ Hz and $Q = 0.25$: solid line, $f_0 = 30$ Hz; thin solid line, $f_0 = 100$ Hz; dashed line, $f_0 = 300$ Hz; and dotted line, $f_0 = 1000$ Hz

Figure 14-3 shows the cascaded bandpass gain and phase response for a fourth-order (M = 2) bandpass Butterworth filter with $f_s = 11025$ Hz and $Q = 0.25$: solid line, $f_0 = 30$ Hz; thin solid line, $f_0 = 100$ Hz; dashed line, $f_0 = 300$ Hz; and dotted line, $f_0 = 1000$ Hz. *Figure 14-4* is identical to *Figure 14-3*, but with $N = 12$ ($M = 6$) and $Q = 1$.

Figures 14-3 and 14-4 were generated using the transfer function H(z) from Equation (14-2b) and the coefficient formulas of Equation (14-3a) through (14-3l). The gain and phase are computed directly from the

Cascaded Bandpass Filter

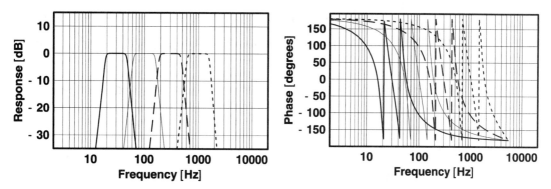

Figure 14-4. Gain and phase response of digital 12th-order (M=6) bandpass Butterworth filter with $f_s = 11025$ Hz and Q = 1: solid line, $f_0 = 30$ Hz; thin solid line, $f_0 = 100$ Hz; dashed line, $f_0 = 300$ Hz; and dotted line, $f_0 = 1000$ Hz

magnitude and argument of H(z) by setting $z = e^{2\pi i f / f_s}$:

$$G(f) = \left| H\left(e^{2\pi i f / f_s}\right) \right| \quad (14\text{-}4a)$$

$$\phi(f) = \frac{180}{\pi} \tan^{-1} \frac{\text{Im}\left[H\left(e^{2\pi i f / f_s}\right)\right]}{\text{Re}\left[H\left(e^{2\pi i f / f_s}\right)\right]} \quad (14\text{-}4b)$$

The steepness of the magnitude response increases with the order of the filter and approaches a straight line on a log-log plot (or *power-law* on a linear-linear plot):

$$G(f) \to \begin{cases} f^{-N/2} & f_0 \ll f \ll f_n \\ f^{N/2} & f \ll f_0 \\ 0 & f \to f_n \end{cases} \quad \varphi(f) \to \begin{cases} -M\frac{\pi}{2} = -N \cdot 45° & f_0 \ll f \leq f_n \\ M\frac{\pi}{2} = N \cdot 45° & f \ll f_0 \end{cases} \quad (14\text{-}5)$$

Equation (14-5) corresponds to the familiar N^{th}-order bandpass filter stopband response approximation of $-3N$ dB/octave or $-10N$ dB/decade when *f* is much less than or much greater than the center frequency, as well as much less than the Nyquist frequency. As *f* approaches the Nyquist frequency, the filter roll-off steepness increases and the gain becomes exactly zero when $f = f_n$. In some ways, this characteristic of

157

DSP Filters

the digital bandpass filter (also true with the low-pass filters) gives an extra boost of stopband suppression as compared to the equivalent analog filter.

Because by convention, as in the analog case, a phase angle is defined to be in the range of −180° to +180°, Equation (14-5) should be modified to reflect this:

$$\varphi(f)_{f \to 0} = \begin{cases} 0 & \text{for } \frac{M}{2} \text{ even} \\ \pi & \text{for } \frac{M}{2} \text{ odd} \end{cases} \quad (14\text{-}6a)$$

$$\varphi(f)_{f \to f_{ny}} = \begin{cases} 0 & \text{for } \frac{M}{2} \text{ even} \\ -\pi & \text{for } \frac{M}{2} \text{ odd} \end{cases} \quad (14\text{-}6b)$$

The gain and phase at the center frequency for all N is:

$$G(f_0) = 1 \quad (14\text{-}7a)$$

$$\phi(f_0) = 0° \quad (14\text{-}7b)$$

As discussed in Chapter 8, the quality factor Q is inversely proportional to the frequency bandwidth Δf of the gain response curve:

$$Q \equiv \frac{f_0}{\Delta f} = \frac{\theta_0}{\Delta \theta} \quad (14\text{-}8)$$

where $\Delta f = f_2 - f_1$ and $\Delta \theta = \theta_2 - \theta_1$. The normalized center frequency θ_0 is related to the normalized edge frequencies θ_1 and θ_2 by:

$$\tan(\theta_0 / 2) = \sqrt{\tan(\theta_1 / 2)\tan(\theta_2 / 2)} \quad (14\text{-}9)$$

where $\theta_0 = 2\pi f_0 / f_s$, $\theta_1 = 2\pi f_1 / f_s$, and $\theta_2 = 2\pi f_2 / f_s$.

Cascaded Bandpass Filter

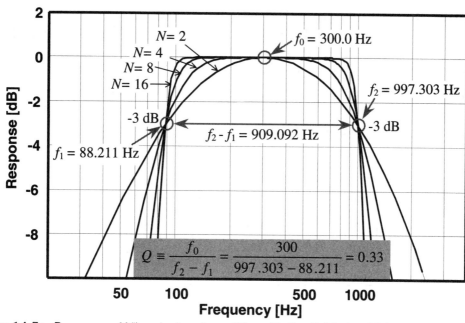

Figure 14-5a. Response of Nth-order bandpass filter with Q = 0.33, f_0 = 300, and f_s = 11025 Hz

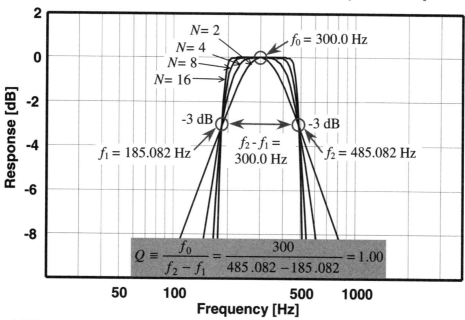

Figure 14-5b. Response of Nth-order bandpass filters with Q = 1.0, f_0 = 300, and f_s = 11025 Hz

DSP Filters

The edge frequencies f_1 and f_2 are computed using an iteractive method described in Chapter 8. *Figure 14-5a* shows the gain response of several N^{th}-order Butterworth bandpass filters with $Q = 0.33$, $f_0 = 300$, and $f_s = 11025$ Hz. Using Equations (14-10a) and (14-10b), the edge frequencies are computed for this case using several values of filter order $N = 2M$. Note that in all cases the response curves intersect at a common point corresponding to the –3 dB gain (or gain = $2^{-1/2}$) at the edge frequencies f_1 and f_2.

$$f_1 = \frac{f_s}{\pi} \tan^{-1}\left\{\frac{\tan^2(\pi f_0 / f_s)}{\tan[(f_1 + f_0/Q)\pi / f_s]}\right\} \qquad (14\text{-}10a)$$

$$f_2 = \frac{f_s}{\pi} \tan^{-1}\left\{\frac{\tan^2(\pi f_0 / f_s)}{\tan[(f_2 - f_0/Q)\pi / f_s]}\right\} \qquad (14\text{-}10b)$$

where,

$$Q \equiv \frac{f_0}{\Delta f} \qquad \Delta f \equiv f_2 - f_1 \qquad \tan^2(\pi f_0 / f_s) = \tan(\pi f_1 / f_s)\tan(\pi f_2 / f_s)$$

Definition of quality factor Q and its relationship to bandwidth Δf of the IIR bandpass filter

As with the second-order bandpass filter, the Q has a finite limit on the low end. There is no theoretical upper limit on Q with the second-order IIR bandpass filter. However, issues concerning round-off errors and other problems with finite size numbers generally lead to a practical upper limit on Q.

$$Q > f_0 / f_n \qquad (14\text{-}11)$$

Q restriction of IIR bandpass filter

Cascaded Bandpass Filter

$$G(f_1) = G(f_2) = \frac{1}{\sqrt{2}} \quad (14\text{-}12a)$$

$$\phi(f_1) = -\phi(f_2) = \begin{cases} \tfrac{1}{2}\pi & \tfrac{M}{2} = 1, 5, 9, \ldots \\ \pi & \tfrac{M}{2} = 2, 6, 10, \ldots \\ -\tfrac{1}{2}\pi & \tfrac{M}{2} = 3, 7, 11, \ldots \\ 0 & \tfrac{M}{2} = 4, 8, 12 \ldots \end{cases} \quad (14\text{-}12b)$$

Gain and phase response of Nth-order IIR bandpass filter at its edge frequencies

Difference Equation

A set of M/2 pair of difference equations can be used to implement the Nth-order Butterworth IIR filter:

$$y_{jk}(n) = 2\left\{\alpha_{jk}\left[x_{jk}(n) - x_{jk}(n-2)\right] + \gamma_{jk}\, y_{jk}(n-1) - \beta_{jk}\, y_{jk}(n-2)\right\} \quad (14\text{-}13)$$

where the input is $x_{11}(n) = x(n)$ and the output is $y(n) = y_{2\frac{M}{2}}(n)$.

Partial C code to implement Equation (14-13) follows below and continues onto the following page:

```
//
// Implementation of Cascaded Bandpass Filter Block,
// class CCascaded_Bandpass
void CCascaded_Bandpass::execute_filter_block_in_place(double *in)
{
    int i;
    double input = *in;
```

DSP Filters

```
        stages[1]->x = input;
        stages[1]->execute_filter_stage();

        for (i= 2; i<= NUM_BANDS; i++)
        {
                stages[i]->x = stages[i-1]->y;
                stages[i]->execute_filter_stage();
        }

        *in = stages[NUM_BANDS]->y;
}


//
// cook_bandpass.cpp
//
// Implementation of a simple Bandpass Filter Stage,
// class CBandpassFilterStage

#include "cook.h"

CBandpassFilterStage::CBandpassFilterStage()
{
        x1 = 0;
        x2 = 0;
        y1 = 0;
        y2 = 0;
}

CBandpassFilterStage::~CBandpassFilterStage()
{
}

void CBandpassFilterStage::execute_filter_stage()
{
        y = 2 * (alpha * (x - x2) + gamma * y1 - beta * y2);
        x2 = x1;
        x1 = x;
        y2 = y1;
        y1 = y;
}
```

15

Cascaded Band-Stop Filter

Digital Filter Network

As shown in Figure *15-1a* and *15-1b*, a digital N^{th}-order band-stop filter can be implemented by cascading the second-order network, which is discussed in Chapter 9, where the second-order transfer function for each of the *M* cascaded sections is expressed as:

$$H_i(z) = \frac{\alpha_i \left(1 - 2\cos\theta_0 z^{-1} + z^{-2}\right)}{\tfrac{1}{2} - \gamma_i z^{-1} + \beta_i z^{-2}} \quad (15\text{-}1)$$

The total transfer function $H(z)$ is the product of the *M* cascaded sections, where the filter order $N = 2M$:

$$H(z) = H_1(z) H_2(z) \cdots H_M(z) \quad (15\text{-}2a)$$

DSP Filters

Instead of a product of M sections as described above, the network can be expressed as M/2 pairs of cascaded sections:

$$H(z) = H_{11}(z) H_{21}(z) H_{12}(z) H_{22}(z) \cdots H_{1\frac{M}{2}}(z) H_{2\frac{M}{2}}(z)$$

$$= \prod_{k=1}^{M/2} \prod_{j=1}^{2} H_{jk}(z) = \prod_{k=1}^{M/2} \prod_{j=1}^{2} \frac{\alpha_{jk}\left(1 - 2\cos\theta_0 z^{-1} + z^{-2}\right)}{\frac{1}{2} - \gamma_{jk} z^{-1} + \beta_{jk} z^{-2}} \quad (15\text{-}2b)$$

where, $\theta_0 \equiv 2\pi f_0 / f_s$. The number of product terms in Equation (15-2b) is still M, but now each pair of $j = 1$, two products correspond to a $k = 1, 2, \ldots M/2$ grouping. The reason for this organization of H(z) may be more obvious by examining Figure 15-2: $j = 1$ corresponds to a second-order band-stop filter whose pole is below f_0, while $j = 2$ corresponds to a band-stop whose pole position is above f_0. Each pair of k sections has a gain value different than one, which is shared by both the $j = 1$ and $j = 2$

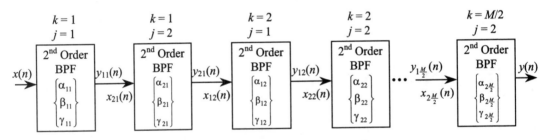

Figure 15-1a. Cascaded N^{th}-order IIR band-stop filter network

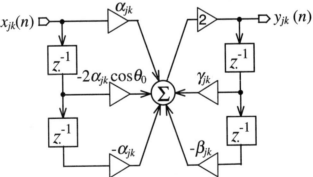

Figure 15-1b. Single second-order section cascaded band-stop filter network

Cascaded Band-Stop Filter

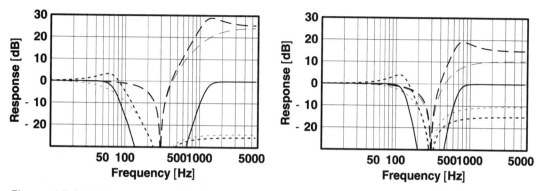

Figure 15-2. Gain response of eighth-order (M=4) Butterworth band-stop filter, showing equivalent gain response of each individual second-order section: f_0 = 300 Hz; left, Q = 0.25; right, Q = 0.5. Solid line, total H(z); dotted line $H_{11}(z)$; dashed line $H_{21}(z)$; thin dotted line $H_{12}(z)$; thin dashed line $H_{22}(z)$

sections. This arrangement leads to a *maximally flat* Butterworth band-stop design. Note that unlike the cascaded bandpass, each band-stop filter section shares a common center frequency θ_0 corresponding to a zero in the complex plane.

The coefficients α_{jk}, β_{jk}, and γ_{jk} in Equation (15-2b) are directly related to the filter Q and center frequency f_0 of the total network. However, like the coefficient formulas of the cascaded bandpass Butterworth filter, the cascaded band-stop coefficient formulas are moderately intricate:

$$\beta_{jk} = \frac{1}{2}\frac{1-\frac{1}{2}d_k \sin\theta_{jk}}{1+\frac{1}{2}d_k \sin\theta_{jk}} \quad (15\text{-}3a)$$

$$\gamma_{jk} = \left(\tfrac{1}{2}+\beta_{jk}\right)\cos\theta_{jk} \quad (15\text{-}3b)$$

$$\alpha_{jk} = \tfrac{1}{2}\left(\tfrac{1}{2}+\beta_{jk}\right)\left(\frac{1-\cos\theta_{jk}}{1-\cos\theta_0}\right) \quad (15\text{-}3c)$$

$$d_k = \sqrt{\frac{d_E D_k}{A_k + \sqrt{A_k^2 - 1}}} \quad (15\text{-}3d)$$

$$d_E = \frac{2\tan\left(\dfrac{\theta_0}{2Q}\right)}{\sin\theta_0} \quad (15\text{-}3e)$$

$$D_k = 2\sin\left(\frac{(2k-1)\pi}{2M}\right) \quad (15\text{-}3f) \qquad A_k = \frac{1+(d_E/2)^2}{D_k\, d_E/2} \quad (15\text{-}3g)$$

$$W_k = B_k + \sqrt{B_k^2 - 1} \quad (15\text{-}3h) \qquad B_k = \frac{D_k}{d_k}(d_E/2) \quad (15\text{-}3i)$$

$$\theta_{1k} = 2\tan^{-1}\left\{\frac{\tan(\theta_0/2)}{W_k}\right\} \quad (15\text{-}3j) \qquad \theta_{2k} = 2\tan^{-1}\{W_k \tan(\theta_0/2)\} \quad (15\text{-}3k)$$

where, $\theta_0 \equiv 2\pi f_0/f_s$.

Figure 15-3 shows the cascaded band-stop gain and phase response for a fourth-order ($M = 2$) band-stop Butterworth filter with $f_s = 11025$ Hz and $Q = 0.25$: solid line, $f_0 = 30$ Hz; thin solid line, $f_0 = 100$ Hz; dashed line, $f_0 = 300$ Hz; and dotted line, $f_0 = 1000$ Hz. *Figure 15-4* is identical to *Figure 15-3*, but with $N = 12$ ($M = 6$).

Figures 15-3 and *15-4* were generated using the transfer function $H(z)$ from Equation (15-2b) and the coefficient formulas of Equation (15-3a) through (15-3l). The gain and phase are computed directly from the *magnitude* and *argument* of $H(z)$ by setting $z = e^{2\pi i f/f_s}$:

$$G(f) = \left| H\left(e^{2\pi i f/f_s}\right) \right| \quad (15\text{-}4a)$$

$$\phi(f) = \frac{180}{\pi}\tan^{-1}\frac{\operatorname{Im}\left[H\left(e^{2\pi i f/f_s}\right)\right]}{\operatorname{Re}\left[H\left(e^{2\pi i f/f_s}\right)\right]} \quad (15\text{-}4b)$$

The steepness of the magnitude response increases with the order of the filter, and approaches a straight line on a log-log plot (or *power-law* on a linear-linear plot):

Cascaded Band-Stop Filter

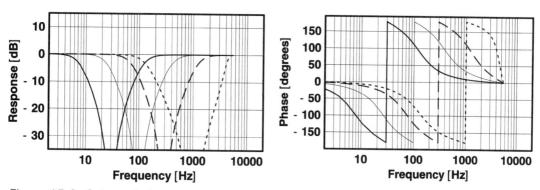

Figure 15-3. Gain and phase response of digital fourth-order (M=2) band-stop Butterworth filter with f_s = 11025 Hz and Q = 0.25: solid line, f_0 = 30 Hz; thin solid line, f_0 = 100 Hz; dashed line, f_0 = 300 Hz; dotted line, f_0 = 1000 Hz

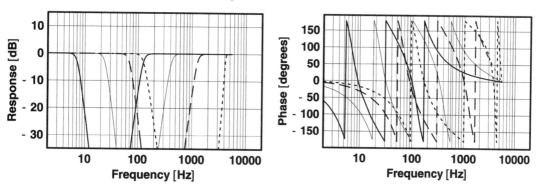

Figure 15-4. Gain and phase response of digital 12th-order (M=6) band-stop Butterworth filter with f_s = 11025 Hz and Q = 0.25: solid line, f_0 = 30 Hz; thin solid line, f_0 = 100 Hz; dashed line, f_0 = 300 Hz; and dotted line, f_0 = 1000 Hz

$$G(f) \to \begin{cases} f^{N/2} & f_0 < f < f_2 \\ f^{-N/2} & f_1 < f < f_0 \\ 1 & f \to f_n \end{cases} \quad \varphi(f) \to \begin{cases} 0° & f_0 \ll f \leq f_n \\ 0° & f \ll f_0 \end{cases} \quad (15\text{-}5)$$

Equation (15-5) corresponds to the familiar N^{th}-order band-stop filter stopband response approximation of $-3N$ dB/octave or $-10N$ dB/decade, when f is close to the center frequency, as well as much less than the Nyquist frequency. The gain approximation in Equation (15-5) becomes more accurate as the filter order N increases and the filter Q decreases (see Figures 15-3 and 15-4).

The behavior of the band-stop phase as $f \to f_0$ depends on the side of the center frequency of the approach:

$$\varphi(f) \underset{f<f_0}{\to} \begin{cases} 0 & \text{for } \frac{M}{2} \text{ even} \\ -\pi & \text{for } \frac{M}{2} \text{ odd} \end{cases} \quad (15\text{-}6a)$$

$$\varphi(f) \underset{f>f_0}{\to} \begin{cases} 0 & \text{for } \frac{M}{2} \text{ even} \\ +\pi & \text{for } \frac{M}{2} \text{ odd} \end{cases} \quad (15\text{-}6b)$$

The gain and phase at the center frequency for all N is:

$$G(f_0) = 0 \quad (15\text{-}7a)$$

$$\phi(f_0) = \text{undefined} \quad (15\text{-}7b)$$

As discussed in Chapter 9, the quality factor Q is inversely proportional to the frequency bandwidth Δf of the gain response curve:

$$Q \equiv \frac{f_0}{\Delta f} = \frac{\theta_0}{\Delta \theta} \quad (15\text{-}8)$$

where $\Delta f = f_2 - f_1$ and $\Delta \theta = \theta_2 - \theta_1$. The normalized center frequency θ_0 is related to the normalized edge frequencies θ_1 and θ_2 by:

$$\tan(\theta_0/2) = \sqrt{\tan(\theta_1/2)\tan(\theta_2/2)} \quad (15\text{-}9)$$

where $\theta_0 = 2\pi f_0 / f_s$, $\theta_1 = 2\pi f_1 / f_s$, and $\theta_2 = 2\pi f_2 / f_s$.

The edge frequencies f_1 and f_2 are computed using an iterative method, which is described in Chapter 9. Figure 15-5a shows the gain response of several N^{th}-order Butterworth band-stop filters with Q = 0.33, f_0 = 300,

Cascaded Band-Stop Filter

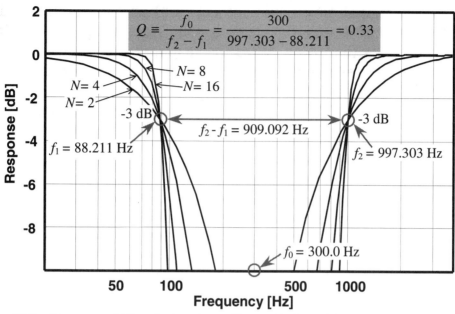

Figure 15-5a. Response of Nth-order band-stop filters with $Q = 0.33$, $f_0 = 300$, and $f_s = 11025$ Hz

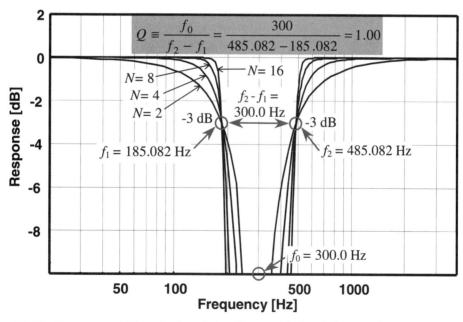

Figure 15-5b. Response of Nth-order band-stop filters with $Q = 1.0$, $f_0 = 300$, and $f_s = 11025$ Hz

DSP Filters

and f_s = 11025 Hz. Using Equations (15-10a) and (15-10b), the edge frequencies are computed for this case using several values of filter order $N = 2M$. Note that in all cases the response curves intersect at a common point corresponding to the −3 dB gain (or gain = $2^{-1/2}$) at the edge frequencies f_1 and f_2.

$$f_1 = \frac{f_s}{\pi} \tan^{-1}\left\{\frac{\tan^2(\pi f_0 / f_s)}{\tan[(f_1 + f_0/Q)\pi / f_s]}\right\} \quad (15\text{-}10\text{a})$$

$$f_2 = \frac{f_s}{\pi} \tan^{-1}\left\{\frac{\tan^2(\pi f_0 / f_s)}{\tan[(f_2 - f_0/Q)\pi / f_s]}\right\} \quad (15\text{-}10\text{b})$$

where,

$$Q \equiv \frac{f_0}{\Delta f} \qquad \Delta f \equiv f_2 - f_1 \qquad \tan^2(\pi f_0 / f_s) = \tan(\pi f_1 / f_s)\tan(\pi f_2 / f_s)$$

Definition of quality factor Q and its relationship to bandwidth Δf of the IIR band-stop filter

As with the second-order band-stop filter, the Q has a finite limit on the low end. There is no theoretical upper limit on Q with the second-order IIR band-stop filter. However, issues concerning round-off errors and other problems with finite size numbers generally lead to a practical upper limit on Q.

$$Q > f_0 / f_{ny} \quad (15\text{-}11)$$

Q restriction of IIR band-stop filter

Cascaded Band-Stop Filter

$$G(f_1) = G(f_2) = \frac{1}{\sqrt{2}} \qquad (15\text{-}12a)$$

$$\phi(f_1) = -\phi(f_2) = \begin{cases} -\frac{1}{2}\pi & \frac{M}{2} = 1, 5, 9, \dots \\ \pi & \frac{M}{2} = 2, 6, 10, \dots \\ \frac{1}{2}\pi & \frac{M}{2} = 3, 7, 11, \dots \\ 0 & \frac{M}{2} = 4, 8, 12 \dots \end{cases} \qquad (15\text{-}12b)$$

Gain and phase response of Nth-order IIR band-stop filter at its edge frequencies

Difference Equation

In order to implement the Nth-order IIR filter, a set of *difference equations* are needed. As previously discussed, the IIR filter implementation is simply a running average of the input and output data for each of the M second-order sections, where M difference equations are used, or in the special case of the band-stop Butterworth filter, M/2 pair of difference equations:

$$y_{jk}(n) = 2\left\{\alpha_{jk}\left[x_{jk}(n) - 2\cos\theta_0\, x_{jk}(n-1) + x_{jk}(n-2)\right] + \gamma_{jk}\, y_{jk}(n-1) - \beta_{jk}\, y_{jk}(n-2)\right\}$$
$$(15\text{-}13)$$

where the input is $x_{11}(n) = x(n)$ and the output is $y(n) = y_{2\frac{M}{2}}(n)$.

Example C code to implement Equation (15-13) follows:

```
//
// Implementation of Cascaded Bandstop Filter Block,
// class CCascaded_Bandstop
void CCascaded_Bandstop::execute_filter_block_in_place(double *in)
```

171

```
{
    int i;
    double input = *in;

    stages[1]->x = input;
    stages[1]->execute_filter_stage();

    for (i= 2; i<= NUM_BANDS; i++)
    {
        stages[i]->x = stages[i-1]->y;
        stages[i]->execute_filter_stage();
    }

    *in = stages[NUM_BANDS]->y;
}
//
// cook_bandstop2.cpp
//
// Implementation of a simple Bandstop2 Filter Stage,
// class CBandstop2FilterStage

#include "cook.h"

CBandstop2FilterStage::CBandstop2FilterStage()
{
    x1 = 0;
    x2 = 0;
    y1 = 0;
    y2 = 0;
}

CBandstop2FilterStage::~CBandstop2FilterStage()
{
}

void CBandstop2FilterStage::execute_filter_stage()
{
    y = 2 * (alpha * (x - 2 * cos_theta_0 * x1 + x2) + gamma * y1 - beta * y2);
    x2 = x1;
    x1 = x;
    y2 = y1;
    y1 = y;
}
```

SECTION II
Filter Projects

16

INTRODUCTION

After studying a subject that is new and unfamiliar, such as digital filters, it is valuable to apply the newly acquired knowledge in some real-life experience or application. This can help solidify understanding of the subject as well as develop confidence in the ability to apply what has been learned. Section II is included to provide that experience and to bridge the gap between developing filter equations performed in previous chapters and their implementation in practical, real-world designs.

Overview

Basic filters were presented in Section I and a good foundation was laid for understanding how to use these filters. Let's apply what has been learned. The following projects do not necessarily focus on developing the most optimal solutions for any particular target application. Rather these projects focus on developing an understanding of how to implement the filters described in Section I with what we hope are interesting and useful audio applications. The projects are presented as implemented in an existing DirectX application. The entire DirectX implementation can be found at www.dspaudiocookbook.com. As a result of being extracted from an existing design, there is a small amount of software overhead included that is not necessarily required for non-DirectX implementations of these filter projects. Care has been taken to clearly indicate where this overhead occurs and to outline what is required for filter implementation.

Section II consists of the following projects:
- Chapter 17: implementing first-order IIR shelving filters to create bass and treble tone controls
- Chapter 18: a simple filter implementation eliminating 60 Hz hum using the second-order IIR band-stop filter
- Chapter 19: implementing a 31-band graphic equalizer using the second-orderIIR peaking filter
- Chapter 20: implementing a 31-band graphic equalizer using the second-order IIR bandpass filter
- Chapter 21: a four-band parametric equalizer using the second-order IIR peaking filters and first-order IIR low-pass and high-pass shelving filters
- Chapter 22: implementing a crossover filter using cascaded second-order IIR low-pass and high-pass filters

Each project provides background to the application, an overview of the system design, and details regarding implementation. System level block diagrams, software flow diagrams, and software implementations in C++ are also provided.

Project Outline

The projects follow a common outline beginning with the introduction. The introduction provides background about the project, including what is to be designed, how the design is normally used, in what applications the design may be applied, and who typically uses such a design. An overview of how it works is also provided here.

Filter Projects

The design requirements outline the specifications and various important details necessary for implementing the project. All filters used in the project are reviewed to provide a convenient reference while going through the project.

The implementations begin with an overview of the system level block diagram. Each component in the block diagram is reviewed except for the user control interface because it can be application specific and not directly related to understanding of the filter design concepts presented in this book. A few notes are provided here regarding the implementation of user interfaces.

User control interfaces may vary greatly from hardware knobs to software GUIs. Regardless of the type of user control interface implemented (hardware or software), it is important to note that the control section must ultimately provide a numerical representation of the control information. For example, in a hardware control interface design a set of analog potentiometers may be implemented that can be multiplexed into an analog-to-digital (A/D) converter. The A/D converter is used to convert the voltage from the analog controls to a digital numeric representation that can be used in calculating coefficients for digital filter implementation. The voltage level controlled by each analog potentiometer is proportional to the variable it represents (gain, center frequency, etc.) with respect to each particular digital filter. The digital output of the A/D converter can be used directly in the calculation of filter coefficients. In a software control interface application, the control positions as displayed in a graphical user interface (GUI) can be read by a software module and applied directly to the coefficient calculation blocks as well. While the user control interface is an important component for any product, it is not relevant to describe the implementation of digital filters and, therefore, will not be reviewed further in the following projects.

Each project concludes with a C++ software solution including software flow diagrams and a detailed description of the software implementation.

DSP Filters

Note that these filters are implemented in the digital domain, so the input analog signal must first be converted to a digital signal via an audio quality A/D converter prior to filtering. Following the filtering process the output digital signal must be converted back to the analog domain via an audio quality digital-to-analog converter (D/A). These are not shown in the project diagrams, but it should be understood that they must exist in the system. Converters are required to provide the audio signals in the digital domain for subsequent processing using the filters described in the following chapters.

Conclusion

Each project is designed to provide an understanding of what audio processor is being created, how it is designed and used in the audio industry, and most importantly, how digital filters are implemented to create the processor. It is our hope that you will find these projects to be helpful in bridging the gap between the cookbook-like equations developed in Section I and their implementation in common digital audio applications.

17

TONE CONTROL

Introduction

This chapter will be a guide, providing step-by-step instructions for implementing a *tone control* using the shelving filters presented in Chapter 11.

Tone controls are equalizers that operate on two bands. *Figure 17-1* displays a typical tone control and its corresponding frequency response. Tone controls provide a quick and easy way to adjust the sound reproduced by an audio system to suit the listener's tastes. The individual bass and treble controls apply a gain or attenuation to the respective frequency bands. The low frequencies are affected only by bass adjustments and high frequencies are affected only by treble adjustments.

Increasing the gain of the bass control increases the relative level of the lower frequencies. This often creates a richer and fuller sound although it can also cause an overly boomy bass when used excessively. Increasing the relative level of the treble control has the effect of bringing out vocals or instruments in the higher frequency ranges. This can allow vocals to be better understood or particular instruments to be heard more clearly. Individuals with excessive hearing loss in the upper

DSP Filters

frequencies may compensate for some of the lost information by increasing the treble gain control as well. However, excessive treble gain creates a tinny or thinner sound for those with normal hearing abilities.

Several common applications such as automobile and home stereo systems, boom boxes, and portable audio players implement tone controls. The design described in this chapter can be used in any of these applications. First-order low-pass and high-pass shelving filters will be utilized to implement tone controls. The respective cutoff frequencies for these filters are fixed depending on the implementation requirements. 100 Hz and 10 kHz are common cutoff frequencies. As shown in *Figure 17-1*, the frequency at which the first-order shelving filters no longer affect the signal is called the hinge point. The low-pass and high-pass hinge points are also fixed depending on the implementation requirements and are not necessarily the same. Often the hinge point for the high-pass shelving filter is lower in frequency than the hinge point for the low-pass shelving filter. This is not a desirable situation as it allows the

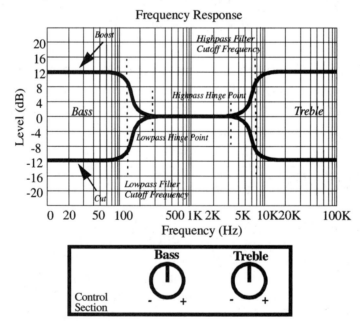

Figure 17-1. Bass and treble tone controls with corresponding frequency response

Tone Control

filters to influence the midband frequencies. Ideally, tone controls have a midband that lies between the bass and treble control bands that is not affected by either control.

As with most other types of equalizers, the application of tone controls has not drastically changed from its inception. However, the implementation has evolved from conventional passive and active analog designs to digital implementations. This chapter provides a solution for implementing a digital tone control.

Design Requirements

The goal of this design is to develop a digital tone control meeting the following specifications:

Specifications:
- Channels: 2 (stereo)
- Bands: Bass with a cutoff frequency at 350 Hz
  Treble with cutoff frequency at 4 kHz
- Gain: +/- 20 dB
- Sample Rate: Variable

This design supports stereo (two channels) audio signals with a bass and treble control applied to both channels. The boost/cut range is +/- 20 dB with controllable step size.

Filter Overview

The tone control is implemented using shelving filters described in Chapter 11. A quick review of the low-pass and high-pass shelving filters is helpful at this point. Equations for implementing a low-pass and high-pass shelving filter as well as the filter realizations are provided here for convenience along with the corresponding frequency and phase response curves.

Low-pass shelving filter overview

A first-order IIR low-pass shelving filter can be implemented by summing the input x(n) with the output of a first-order low-pass filter scaled by µ-1, as shown in *Figure 17-2*. The low-pass output scale factor is chosen so that when µ = 1, the output is equal to the input, y(n) = x(n). The coefficient depends on the shelving level g, as $\mu \equiv 10^{g/20}$, where g is the boost/cut gain in dB.

The frequency and phase response of the low-pass shelving filter is displayed in *Figure 17-3*. Note that since the network is first order, the phase excursion will not exceed ±90°. The maximum and minimum phase value is controlled by the shelving filter's boost/cut gain factor g.

The difference equation describing a low-pass shelving filter is provided in Equations (17-1a) and (17-1b) while the coefficients α and γ are determined by Equation (17-1c).

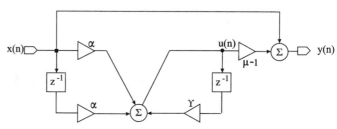

Figure 17-2. IIR low-pass shelving filter

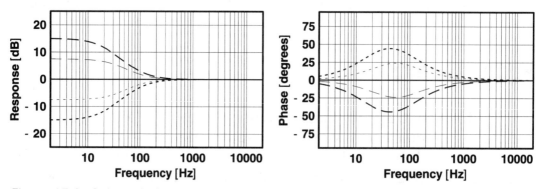

Figure 17-3. Gain and phase of low-pass shelving filter with f_c = 30 and f_s = 44100 Hz: dotted line, g = -15 dB; thin dotted line, g = -7.5 dB; solid line, g = 0 dB; thin dashed line, g = +7.5 dB; and dashed line, g = +15 dB

$$u(n) = \alpha\left[x(n) + x(n-1)\right] + \gamma\, u(n-1) \qquad (17\text{-}1a)$$

$$y(n) = x(n) + (\mu - 1)\, u(n) \qquad (17\text{-}1b)$$

$$\gamma = \frac{1 - \left(\dfrac{4}{1+\mu}\right)\tan\dfrac{\theta_c}{2}}{1 + \left(\dfrac{4}{1+\mu}\right)\tan\dfrac{\theta_c}{2}} \qquad \alpha = (1-\gamma)/2 \qquad (17\text{-}1c)$$

where, $\theta \equiv 2\pi f / f_s$ $\qquad \theta_c \equiv 2\pi f_c / f_s$

Difference equation of digital first-order low-pass shelving filter with coefficient formulas

High-pass shelving filter overview

A first-order IIR high-pass shelving filter can be implemented by summing the input *x(n)* with the output of a first-order high-pass filter scaled by *μ*-1, as shown in *Figure 17-4*. The high-pass output scale factor is chosen so that when *μ* = 1, the output is equal to the input, *y(n)* = *x(n)*. The coefficient depends on the shelving level g, as $\mu \equiv 10^{g/20}$, where g is the boost/cut gain in dB.

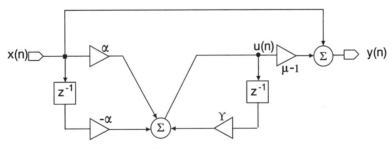

Figure 17-4. IIR High-pass Shelving Filter

DSP Filters

The frequency and phase response of the high-pass shelving filter is displayed in *Figure 17-5*. Note that since the network is first order, the phase excursion will not exceed ±90°. The maximum and minimum phase value is controlled by the shelving filter boost/cut gain factor g.

The difference equation describing a high-pass shelving filter is provided in Equations (17-2a) and (17-2b), while the coefficients α and γ are determined by Equation 17-2c.

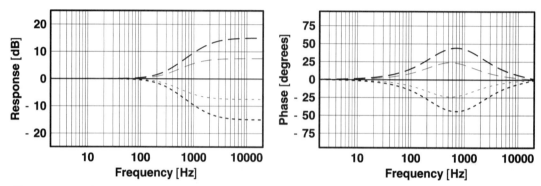

Figure 17-5. Gain and phase of high-pass shelving filter with f_c = 1000 and f_s = 44100 Hz: dotted line, g = -15 dB; thin dotted line, g = -7.5 dB; solid line, g = 0 dB; thin dashed line, g = +7.5 dB; dashed line, g = +15 dB

$$u(n) = \alpha\left[x(n) - x(n-1)\right] + \gamma\, u(n-1) \qquad (17\text{-}2\text{a})$$

$$y(n) = x(n) + (\mu - 1)\, u(n) \qquad (17\text{-}2\text{b})$$

$$\gamma = \frac{1 - \left(\dfrac{1+\mu}{4}\right)\tan\dfrac{\theta_c}{2}}{1 + \left(\dfrac{1+\mu}{4}\right)\tan\dfrac{\theta_c}{2}} \qquad \alpha = (1+\gamma)/2 \qquad (17\text{-}2\text{c})$$

$$\text{where,}\quad \theta \equiv 2\pi f / f_s \qquad \theta_c \equiv 2\pi f_c / f_s$$

Difference equation of digital first-order high-pass shelving filter with coefficient formulas

Tone Control

Functional Block Diagram

The tone control's functional block diagram is shown in *Figure 17-6*. The bass and treble controls provide gain information to the respective coefficient calculation blocks. These blocks determine the appropriate coefficients α and γ to be applied to the respective shelving filters (f_1 and f_2). f_1 corresponds to the low-pass shelving filter, while f_2 represents the high-pass shelving filter. The filters are cascaded together and operate on the input digital signal. The digital output is the processed audio signal corresponding to the user's taste as determined by the bass and treble control settings.

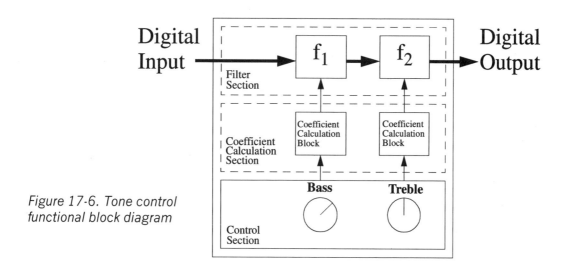

Figure 17-6. Tone control functional block diagram

Control section

The control section of the tone control can be implemented in either hardware or in software. Regardless of the type of design required (hardware or software), the control section must provide a numerical representation of the bass and treble control positions.

185

DSP Filters

Coefficient calculation section

The coefficient calculation blocks receive the bass and treble gain values from the tone controls. The gain values are provided to the coefficient calculation blocks in digital form as numbers representing the respective gains. Using these bass and treble gain values, the coefficient calculation blocks determine the α and γ coefficients required to correctly implement the corresponding low-pass and high-pass shelving filters. The coefficients are determined using the difference equations from Chapter 11 and repeated in this chapter as Equations (17-1a) and (17-2a).

Filter Section

The filter blocks f_1 and f_2 in the filter section apply the coefficients *a* and *g* to the respective low-pass and high-pass shelving filters. The filters are cascaded as shown in *Figure 17-6* and applied to the input digital signal. The filter section effectively modifies the signal frequency response based on the bass and treble control settings.

Flow Diagram Descriptions

The following describes the implementation of the coefficient calculation and filter sections outlined in the block diagram in *Figure 17-6*. Both the coefficient calculation and filter sections are described. The control section is not overviewed as it can be implemented in many ways in either hardware or software and does not directly effect how the filter equations are utilized in the implementation of the tone controls. The coefficient calculation and filter sections operate asynchronously and are described in more detail in the sections that follow.

Figures 17-7 through *17-10* display flow diagrams utilized for implementing the coefficient calculation section. As shown in *Figure 17-7*, there are three system level variables that effect the coefficients α and γ required to implement the tone controls. The *sample rate (fs), shelving filter*

Tone Control

cutoff frequency (f_0), and gain factor (g) each affect the resulting coefficients used to implement the tone control. The coefficients change dynamically with changes to any of these variables.

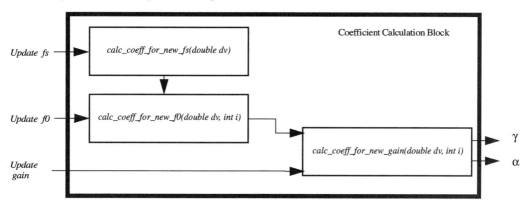

Figure 17-7. Coefficient calculation block diagram

Updating the sample rate

Figure 17-8 shows the software flow diagram implementing the coefficient changes based on a change in sample rate (f_s). When the sample rate (f_s) changes, the *maximum cutoff frequency (max_f_0)* is updated. *max_f_0* must be less than *fs/2*, which is the Nyquist frequency. This is performed in the *calc_coeff_for_new_f$_s$(double dv)* routine. The normalized cutoff frequency (θ) is dependent upon the f_s and must also be updated for each high-pass and low-pass shelving filters. This is updated

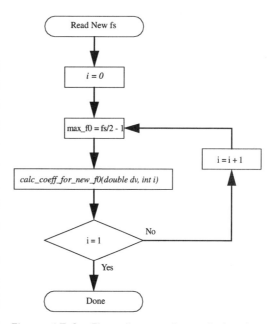

Figure 17-8. Flow diagram for updating f_s Routine: calc_coeff_for_new_fs(double dv)

DSP Filters

by the routine, which is called from *calc_coeff_for_new_f₀(double dv, int i)*. The coefficients α and γ can then be calculated for each filter from the updated θ.

Updating the cutoff frequency

As shown in *Figure 17-9*, when the cutoff frequency (f_o) is adjusted there are several parameters to evaluate. The cutoff frequency (f_o) cannot excede the Nyquist frequency noted here as the maximum cutoff frequency (*max_f_o*). There is also a *minimum cutoff frequency* (*min_f_o*) that should be evaluated if, for example, it is desireable to match a speaker's frequency range. Note that this parameter is not normally adjustable in tone controls. User controls are not typically provided to adjust these parameters, but in this design we have made it adjustable and the system designer can choose to extend the control to the user or not to give the user controls for cutoff frequency. The normalized cutoff frequency θ is dependent upon both the cutoff frequency and sample rate and thus must be recalculated for the corresponding filter whenever either of these variables changes.

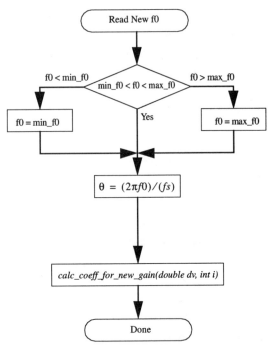

Figure 17-9. Flow diagram for updating f0 Routine:
calc_coeff_for_new_f0(double dv, int i)

Updating the gain

The third variable, gain (g), also affects the coefficients for the respective shelving filter. This is shown in *Figure 17-10*. The gain is com-

Tone Control

pared to the system designer's maximum *(max_gain)* and minimum gain *(min_gain)* values. These may be established based on available headroom or noise floor considerations. The gain factor is used to calculate the variable μ. α and γ can then be calculated using μ.

Filtering section

The filter section flow diagrams are shown in *Figures 17-11* through *17-13*. The updating of the coefficients in the coefficient calculation blocks can occur asynchronously with the processing of the samples in the filter section. *Figure 17-11* displays the flow diagram, which involves reading the samples from the input stream

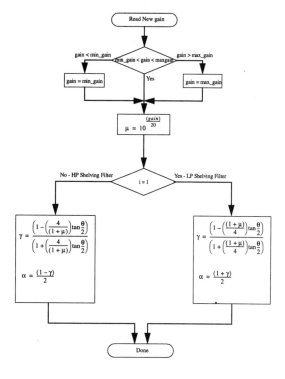

Figure 17-10. Flow diagram for updating gain Routine: calc_coeff_for_new_gain(double dv, int i)

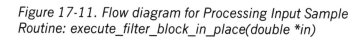

Figure 17-11. Flow diagram for Processing Input Sample Routine: execute_filter_block_in_place(double *in)

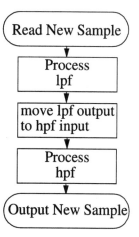

DSP Filters

where the low-pass shelving filter operates on the sample. The output from the low-pass shelving filter is input to the high-pass shelving filter in cascade. The high-pass shelving filter processes the sample and outputs the result to the rest of the system.

The low-pass shelving filter processes the input sample as shown in the flow diagram of *Figure 17-12*. The low-pass section of the filter is

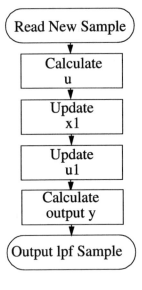

Figure 17-12. Flow diagram for low-pass shelving filter Implementation Routine: stage[0]->execute_filter_stage()

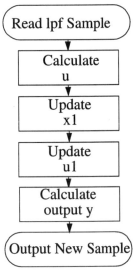

Figure 17-13. Flow diagram for high-pass shelving filter Implementation Routine: stage[1]>execute_filter_stage()

Tone Control

calculated resulting in the output scale factor *u* (not to be confused with *µ*—the gain factor). The delayed input sample *x1* is updated, the delayed output sample *u1* is updated, and the low-pass shelving filter output is calculated as *y*.

The high-pass shelving filter operates similar to the low-pass shelving filter as shown in *Figure 17-13*.

Software Description

All filter-related information is stored in a matrix as shown in Table 17-1. This matrix stores the variables required for processing each of the shelving filters. Column 0 contains variables for the low-pass shelving filter while column 1 contains variables for the high-pass shelving filter. Each column is referenced in the software by the index *i* as a part of an array *stages[i]*. Table 17-1 displays the tone control variables in their initialized states.

Table 17-1. Shelving Filter variable matrix: stages[i]->variable

Variable	*i* ->	0	1
min_f0 - Minimum LP Shelving Filter Cutoff		0.1	0.1
f0 - LP Shelving Filter Cutoff		300	4000
max_f0 - Maximum LP Shelving Filter Cutoff		22049	22049
min_gain - Minimum Filter gain		-20	-20
Gain - Filter gain		0	0
max_gain - Maximum Filter gain		20	20
min_Q - Minimum Filter Quality factor		0.1	0.1
Q - Filter Quality factor		1	1
max_Q - Maximum Quality factor		20	20
x - Filter Input Sample		0	0
x1 - Peaking Filter Input Sample		0	0
y - Filter Output Sample		0	0
u1 - Previous Filter Output Sample		0	0
mu - $10^{G/20}$		1	1
theta0 - $2*pi*f_0/f_s$		.00306	.00385
gamma - Filter feedback coefficient		-.6277	-.6277
alpha - Filter feed-forward coefficient		-.6277	-.6277

See also Figure 17-2, which displays the low-pass shelving filter

DSP Filters

The software routines described below implement the functions outlined by the flow diagrams. There are several routines dedicated to initialization of the array as well. The following describes the initialization, coefficient calculation section, and filtering section. Each line of the software is numbered to assist in the description of operation.

Initialize the variable matrix

The implementation begins with a description of the initialization routines required to establish the variable matrix in Table 1 to a known stable state. Initialization is performed prior to activation of the tone control. It is performed only once prior to activation or following any system reset condition. A line-by-line analysis of the initialization routine is provided here.

```
// Implementation of Simple Shelving Tone Control Filter Block,


1  const double TWO_PI = 2.0 *3.14159265358979323846;
2  const double PI = 3.14159265358979323846;
3  const int NUM_BANDS = 2;
4  CShelving_Tone::CShelving_Tone()
5  {
6       stages = new (CCookFilterStage * [NUM_BANDS]);
7       stages[0] = new CLP_ShelvingFilterStage;
8       stages[1] = new CHP_ShelvingFilterStage;

         // Initialization of LP filter
9       {
10          stages[0]->min_f0 = 0.1;
11          stages[0]->min_gain = -20.0;  //db
12          stages[0]->max_gain = +20.0;
13          stages[0]->min_Q = 0.1;
14          stages[0]->max_Q = 20.0;
15          stages[0]->gain = 0.0;
16          stages[0]->Q =1.0;
```

Tone Control

```
17         stages[0]->f0 = 300.0;
18         stages[0]->max_f0 = 22049.0;

// Initialization of HP filter
19         {
20                 stages[1]->min_f0 = 0.1;
21                 stages[1]->min_gain = -20.0;  //db
22                 stages[1]->max_gain = +20.0;
23                 stages[1]->min_Q = 0.1;
24                 stages[1]->max_Q = 20.0;
25                 stages[1]->gain = 0.0;
26                 stages[1]->Q =1.0;
27                 stages[1]->f0 = 4000.0;
28                 stages[1]->max_f0 = 22049.0;
29         }
30         num_filter_stages = NUM_BANDS;
31         enabled = 1;
32 }
```

- Lines 1, 2, and 3 define constants "2pi" ($2*\pi$) and "pi" (π) as well as the number of filter bands (2 – bass and treble).

- Lines 4, 5, and 32 define the boundaries of this routine. Everything between these lines constitutes the main initialization routine.

- Line 6 calls a routine that clears the memory used for the variable matrix.

- Lines 7 and 8 invoke constructors that establish memory in the system for the matrix variables and associates the variables with the rows of the matrix. These routines are shown below.

- Lines 10 through 19 initialize the low-pass shelving filter variables in the matrix as shown.

    Line 10 initializes the minimum cutoff frequency for the low-pass shelving filter.

    Line 11 initializes the minimum gain for the low-pass shelving filter.

    Line 12 initializes the maximum gain for the low-pass shelving filter.

    Line 13 initializes the minimum Q for the low-pass shelving filter. Although Q is not a controlled low-pass filter variable, it is required overhead in our DirectX control interface.

DSP Filters

> *Line 14 initializes the maximum Q for the low-pass shelving filter. Although Q is not a controlled low-pass filter variable, it is required overhead in our DirectX control interface.*
>
> *Line 15 initializes the gain for the low-pass shelving filter.*
>
> *Line 16 initializes the Q for the low-pass shelving filter. Although Q is not a controlled low-pass filter variable, it is required overhead in our DirectX control interface.*
>
> *Line 17 initializes the cutoff frequency for the low-pass shelving filter.*
>
> *Line 18 initializes the maximum cutoff frequency for the low-pass shelving filter.*

- **Lines 19 through 29 initialize the high-pass filter variables in the matrix as shown.**

  > *Line 20 initializes the minimum cutoff frequency for the high-pass shelving filter.*
  >
  > *Line 21 initializes the minimum gain for the high-pass shelving filter.*
  >
  > *Line 22 initializes the maximum gain for the high-pass shelving filter.*
  >
  > *Line 23 initializes the minimum Q for the high-pass shelving filter. Although Q is not a controlled high-pass filter variable, it is required overhead in our DirectX control interface.*
  >
  > *Line 24 initializes the maximum Q for the high-pass shelving filter. Although Q is not a controlled high-pass filter variable, it is required overhead in our DirectX control interface.*
  >
  > *Line 25 initializes the gain for the high-pass shelving filter.*
  >
  > *Line 26 initializes the Q for the high-pass shelving filter. Although Q is not a controlled high-pass filter variable, it is overhead required in our DirectX control interface.*
  >
  > *Line 27 initializes the cutoff frequency for the high-pass shelving filter.*
  >
  > *Line 28 initializes the maximum cutoff frequency for the high-pass shelving filter.*

- **Line 30 establishes num_filter_stages equal to NUM_BANDS, which in this case is 2.**

- **Line 31 initializes the variable enabled, which indicates that initialization has been completed and the tone control can now be enabled.**

Tone Control

The following routine is the constructor used in the initialization process and initializes the low-pass shelving filter state variables called in the initialization routine.

```
1  CLP_ShelvingFilterStage::CLP_ShelvingFilterStage()
2  {
3      x = 0;
4      y = 0;
5      x1 = 0;
6      u1 = 0;
7  }
```

The following routine is the constructor used in the initialization process and initializes the high-pass shelving filter state variables called in the above initialization routine.

```
1  CHP_ShelvingFilterStage::CHP_ShelvingFilterStage()
2  {
3      x = 0;
4      y = 0;
5      x1 = 0;
6      u1 = 0;
7  }
```

Following the completion of the initialization routine, the data contained in the matrix is as shown in Table 1-17.

Coefficient calculation section

Whenever a parameter is adjusted for one or both of the filters such as sample rate (f_s), cutoff frequency (f_o), or gain (*g*) the coefficient calculation blocks calculate new coefficients. The flow diagrams shown in *Figures 17-8, 17-9,* and *17-10* describe the operation of the coefficient calculation blocks. The following routines implement these flow diagrams. These routines are designed to work together such that they do not duplicate operations. For example, when the sample rate changes the rou-

DSP Filters

tine *calc_coeff_for_new_f$_s$ (double dv)* is called. It performs functions unique to sample rate change and then calls the routine *calc_coeff_for_new_f$_0$ (double dv, int i)* that will then call the routine *calc_coeff_for_new_gain (double dv, int i)* to complete the necessary changes. *calc_coeff_for_new_f$_0$ (double dv, int i)* performs operations related to f_0 and f_s while *calc_coeff_for_new_gain (double dv, int i))* performs operations related to f_0, f_s and *gain*.

```
1  void CShelving_Tone::calc_coeff_for_new_fs (double dv)
2  {
3      int i;
4      f_s = dv;
5      for (i = 0; i< NUM_BANDS; i++)
6      {
7          stages[i]->max_f_0 = f_s/2.0 -1.0;
8          calc_coeff_for_new_f_0 (stages[i]->f0, i);
9      }
10 }
```

- Lines 1, 2, and 10 define the boundaries of this routine. Everything between these lines constitutes the *calc_coeff_for_new_f$_s$ (double dv)* routine.

- Line 3 defines the index *i*.

- Line 4 establishes the variable f_s to be equal to the value of the passed parameter *dv* in Hertz.

- Lines 5, 6, and 9 define a loop that recalculates coefficients based on a sample rate change. This loop operates twice, once for the low-pass shelving filter (column 0) and once for the high-pass shelving filter (column 1).

- Line 7 calculates the maximum cutoff frequency value base on Nyquist's theorem. The maximum cutoff frequency should not exceed one-half of the sample rate frequency.

Tone Control

- Line 8 calls the routine calc_coeff_for_new_f_0 (stages[i]->f_0, i) shown below, which will pass the existing cutoff frequency f0 and the index i indicating which filter to process. The calc_coeff_for_new_f_0(double dv, int i) routine will calculate θ for each filter based on the updated sample rate and subsequently update the respective coefficients α and γ.

Typically the cutoff frequency in a tone control is constant, but in the event the implementation requires a variable cutoff frequency the following routine can be used to make the appropriate coefficient changes. This routine is called either when one of the filter's cutoff frequencies has been altered or from the calc_coeff_for_new_f_s (double dv) routine. The routine calc_coeff_for_new_f_0 (double dv, int i) will operate on either filter (high-pass shelving or low-pass shelving) as defined by the index i passed to it. The value dv passed to the routine is the cutoff frequency in hertz. The following describes the operation of this routine.

```
1 void CShelving_Tone::calc_coeff_for_new_f0 (double dv, int i)
2 {
3      CCookFilterStage *st = stages[i];
4      if (dv > st->max_f0) dv = st->max_f0;
5      else if (dv < st->min_f0) dv = st->min_f0;
6      st->f0 = dv;
7      st->theta0 = TWO_PI * st->f0/fs;
8      calc_coeff_for_new_gain (st->gain, i);
9 }
```

- Lines 1, 2, and 9 define the boundaries of this routine. Everything between these lines constitutes the calc_coeff_for_new_f_0 (double dv, int i) routine.

- Line 3 establishes st as a pointer to the variables in column i of the variables matrix. If the value 0 is passed to the routine as the index i then st points to the variables in column 0 of the matrix (low-pass shelving filter). If the value 1 is passed to the routine as the index i then st points to the variables in column 1 of the matrix (high-pass shelving filter).

DSP Filters

- Line 4 tests the cutoff frequency *dv* to verify that it does not exceed the maximum cutoff frequency (*max_f$_0$*) established in the column pointed to by *st*. If the cutoff frequency exceeds the maximum value then it is set equal to the maximum cutoff frequency.
- Line 5 tests the cutoff frequency *dv* to verify that it is not less than the minimum cutoff frequency boundaries (*min_f$_0$*) established in the column pointed to by *st*. If the cutoff frequency is less than the minimum value then it is set equal to the minimum cutoff frequency.
- Line 6 sets the variable *f$_0$* in the column pointed to by *st* of the matrix equal to the cutoff frequency value passed to the routine *dv*.
- Line 7 calculates the normalized cutoff frequency θ in the column pointed to by *st* of the matrix given both the cutoff frequency *f0* in the column pointed to by *st* and the sample rate *fs*.
- Line 8 calls the routine *calc_coeff_for_new_gain (st->gain, i)*. The original gain value pointed to by *st* is passed along with the index *i* indicating which filter is to be processed.

Coefficients α and γ are modified only by the *calc_coeff_for_new_gain (double dv, int i)* routine below. This routine can be called by directly altering the bass or treble gain or from either of the previous two routines *calc_coeff_for_new_f$_s$ (double dv)* and *calc_coeff_for_new_f$_0$ (double dv, int i)*. In the former case, the new gain is passed to the routine as the variable *dv* along with the filter designation denoted by the index *i*. This routine will be called if any updates are to be made to the coefficients due to changes in sample rate, cutoff frequency, or gain.

```
1 void CShelving_Tone::calc_coeff_for_new_gain (double dv, int i)
2 {
3     CCookFilterStage *st = stages[i];

4     if (dv > st->max_gain) dv = st->max_gain;
5     else if (dv < st->min_gain) dv = st->min_gain;
```

Tone Control

```
6      st->gain = dv;
7      st->mu = pow(10.0,st->gain/20.0);

8      if (i == 1 ) // HP shelving filter
9      {
10         st->gamma = (1.0-(((1.0+st->mu)/4.0) * tan(st->theta0/2.0)))/
                       (1.0+(((1.0+st->mu)/4.0) * tan(st->theta0/2.0)));
11         st->alpha = (1.0 + st->gamma)/2.0;
12     }
13     else // LP shelving
14     {
15         st->gamma = (1.0-((4.0/(1.0+st->mu)) * tan(st->theta0/2.0)))/
                       (1.0+((4.0/(1.0+st->mu)) * tan(st->theta0/2.0)));
16         st->alpha = (1.0 - st->gamma)/2.0;
17     }
18 }
```

- Lines 1, 2, and 18 define the boundaries of this routine. Everything between these lines constitutes the *calc_coeff_for_new_gain (double dv, int i)* routine.

- Line 3 establishes *st* as a pointer to the variables in column *i* of the variables matrix. If the value 0 is passed to the routine as the index *i* then *st* points to the variables in column 0 (low-pass shelving filter) of the matrix. If the value 1 is passed to the routine as the index *i* then *st* points to the variables in column 1 (high-pass shelving filter) of the matrix.

- Line 4 tests the gain *dv* to verify that it does not exceed the maximum gain (*max_gain*) established in the column pointed to by *st*. If the gain exceeds the maximum gain value then it is set equal to the maximum gain.

- Line 5 tests the gain *dv* to verify that it is not less than the minimum gain boundaries (*min_gain*) established in the column pointed to by *st*. If the gain is less than the minimum gain value then it is set equal to the minimum gain.

DSP Filters

- Line 6 sets the variable *gain* in the column pointed to by *st* of the matrix equal to the gain value passed to the routine *dv*.

- Line 7 calculates μ in the column pointed to by *st* given the gain in the column pointed to by *st* as $10^{(g/20)}$.

- Lines 8 through 12 determine the coefficients α and γ for the high-pass shelving filter when the index *i* is 1 indicating operation using variables in column 1.

    Lines 8, 9, and 12 define the boundaries of the if statement given that i equals 1.

- Lines 10 and 11 calculate α and γ respectively based on Equation 17-2c.

- Lines 13 through 17 determine the coefficients α and γ for the low-pass shelving filter when the index *i* is 0 indicating operation using variables in column 0.

    Lines 13, 14, and 17 define the boundaries of the if statement given that i equals 0.

    Line 15 and 16 calculates a and g respectively based on Equation 17-1c.

Filter Section

The filter section operates asynchronously with the coefficient calculation section. Each sample is processed as it is streamed into the filter section. The filter section is implemented in the routine *execute_filter_block_in_place(double *in)*. This routine is called for each sample processed in the filter section. A description of this routine is provided on the following page.

```
1  void CShelving_Tone::execute_filter_block_in_place(double *in)
2  {
3      int i;
4      double input = *in;

5      stages[0]->x = input;
6      stages[0]->execute_filter_stage();
7      stages[1]->x = stages[0]->y;
8      stages[1]->execute_filter_stage();

9      *in = stages[1]->y;
10 }
```

Tone Control

- Lines 1, 2, and 10 define the boundaries of this routine. Everything between these lines constitutes the *execute_filter_block_in_place(double *in)* routine.

- Line 3 defines the index *i*.

- Line 4 defines *input* to be a value equal to the value of the sample pointed to by **in*. **in* points to the next sample to be processed by the filter section.

- Line 5 copies the sample (*input*) to the matrix variable *x* location in column 0 (low-pass shelving filter).

- Line 6 calls the routine *execute_filter_stage()* which processes the input sample using the low-pass shelving filter as noted by index *i*=0. All processing parameters for *execute_filter_stage()* will come from column 0 of the matrix.

- Line 7 copies the low-pass filter's output sample *stages[0]->y* (where y denotes the output sample and 0 refers to the matrix column) to the high-pass filter's input sample *stages[1]->x* (where x denotes the input sample and 1 refers to the matrix column).

- Line 8 calls the routine *execute_filter_stage()* which processes the sample using the high-pass shelving filter as noted by index *i*=1. All processing parameters for *execute_filter_stage()* will come from column 1 of the matrix.

- Line 9 stores the output sample of the high-pass shelving filter, *stages[1]->y* (where y denotes the output sample and 1 refers to the matrix column) in the memory location pointed to by **in*.

The pointer **in* is then used by an output routine to pass the sample on for further processing or to an output D/A converter.

DSP Filters

The *execute_filter_stage()* implements the second-order IIR shelving filter as shown in network diagram in *Figures 17-2* and *17-4*. The coefficients and state variables are stored in the respective columns of the matrix for each filter. The variables utilized in this routine are listed below:

```
x  - Input sample
u  - lowpass filter Output sample
x1 - past input sample
u1 - past lowpass filter output sample
alpha, gamma - filter coefficients
mu - gain coefficient
```

This routine implements the low-pass shelving filter difference equation, Equation (17-1a), and the flow diagram shown in *Figure 17-12*. The following describes this routine.

```
// Implementation of a simple LP_Shelving Filter Stage,
// class CLP_ShelvingFilterStage

1  void CLP_ShelvingFilterStage::execute_filter_stage()
2  {
3      double u;

4      u = (alpha * (x + x1) + (gamma * u1));
5      x1 = x;
6      u1 = u;
7      y = (u * (mu - 1.0)) + x;
8  }
```

- Lines 1, 2, and 8 define the boundaries of this routine.
- Line 3 defines the variable u.
- Line 4 calculates the low-pass filter output value u.
- Lines 5 and 6 update the filter state variables (x1, u1) for the low-pass shelving filter.
- Line 7 calculates the low-pass shelving filter's output value and stores it in the variable y.

Tone Control

This routine implements the high-pass shelving filter difference equation, Equation (17-2a), and the flow diagram shown in *Figure 17-12*. The following describes this routine.

```
// Implementation of a simple HP_Shelving Filter Stage,
// class CHP_ShelvingFilterStage

1 void CHP_ShelvingFilterStage::execute_filter_stage()
2 {
3      double u;
4      u = (alpha * (x - x1) + (gamma * u1) );
5      x1 = x;
6      u1 = u;
7      y = (u * (mu - 1.0)) + x;
8 }
```

- Lines 1, 2, and 8 define the boundaries of this routine.
- Line 3 defines the variable u.
- Line 4 calculates the high-pass filter output value u.
- Lines 5 and 6 update the filter state variables (x1, u1) for the high-pass shelving filter.
- Line 7 calculates the high-pass shelving filter's output value and stores it in the variable y.

Disabling the tone control in a DirectX application requires a destructor routine that gives up memory to the rest of the system. Embedded applications on dedicated hardware do not require this routine. It is provided here for completeness. The *~CShelving_Tone()* routine clears the memory used for the matrix and gives it back to the system.

```
1 CShelving_Tone::~CShelving_Tone()
2 {
3      int i;
4      if (stages)
```

```
5       {
6               for (i = 0; i< NUM_BANDS; i++)
7               {
8                       delete stages[i];
9               }
10              delete stages;
11              stages = 0;
12      }
13 }
```

18
60 Hz
Hum Eliminator

Introduction

There are several potential sources of noise in professional and home consumer audio equipment, but one type of noise is more common than others—the 60 Hz hum. This noise is created by poor power supplies, transformers, or electromagnetic interference (EMI) sourced by the main power supply and is characterized by a 60 Hz sine wave and its harmonics interjected into the audio signal. In European countries and various other locations throughout the world it is actually 50 Hz and related harmonics. Live performances are susceptible to 60 Hz hum as many venues do not provide an adequate power source for all of the various sound, light, and communications equipment. There are often ways to resolve these noise issues without signal processing, but when other means are not successful, it is helpful to have a hum eliminator available

DSP Filters

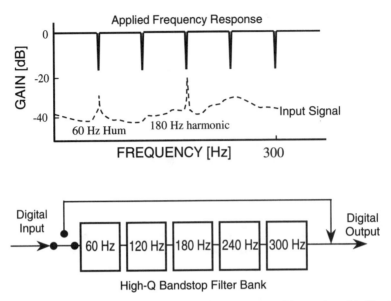

Figure 18-1. Hum eliminator witih five cascaded band-stop filters: $f_1 = 60$ Hz; $f_2 = 120$ Hz; $f_3 = 180$ Hz; $f_4 = 240$ Hz; $f_5 = 300$ Hz

to quickly and easily remove the annoying hum. The dashed line of *Figure 18-1* shows the frequency plot of a signal with degradation due to 60 Hz and its third harmonic.

Design Requirements

The goal of this design is to develop a two-channel 60 Hz hum eliminator applying a notch filter at 60 Hz and closely related harmonics.

Specifications:
    Channels: 2(stereo)
    Harmonics Removed: 60 Hz (fundamental)
                         120 Hz (second Harmonic)
                         180 Hz (third Harmonic)
                         240 Hz (fourth Harmonic)
                         300 Hz (fifth Harmonic)

60 Hz Hum Eliminator

This design will support stereo (two channels) audio signals with independent processing for each channel having five bands at multiples of 60 Hz. Each band-stop filter will use a high Q factor for minimal degradation of the signal outside of the *notch* frequencies. The cut range, in theory, is infinite, but due to the arithmetic of finite numbers, the actual cut may be finite. For 16-bit numbers, the attenuation at the notch frequencies should be expected to be better than –90 dB.

Band-Stop Filter Overview

Recall from Chapter 9 that a second-order IIR band-stop filter can be implemented by performing a running average on the input sample $x(n)$, two previous input samples, plus two previous output samples. The difference equations and coefficient

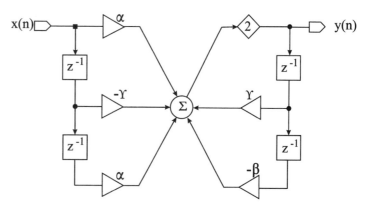

Figure 18-2. Digital band-stop filter network

formulas from Chapter 9 are repeated below as Equations (18-1a) and (18-1b). The normalized center frequency is $\theta_0 = 2\pi f_0 / f_s$, where the notch frequency is f_0. The center frequency f_0, sample rate f_s, and quality factor Q, will each influence the calculation of the coefficients β, γ, and α, which are computed according to Equation (18-1b). The coefficients are then utilized to compute the difference equation in Equation (18-1a) and calculate the output sample $y(n)$.

DSP Filters

$$y(n) = 2\{\alpha\, x(n) - \gamma\, x(n-1) + \alpha\, x(n-2) + \gamma\, y(n-1) - \beta\, y(n-2)\} \quad (18\text{-}1a)$$

$$\beta = \frac{1}{2}\frac{1-\tan[\theta_0/(2Q)]}{1+\tan[\theta_0/(2Q)]} \qquad \gamma = \left(\tfrac{1}{2}+\beta\right)\cos\theta_0 \qquad \alpha = \left(\tfrac{1}{2}+\beta\right)/2 \quad (18\text{-}1b)$$

where, $\theta_0 \equiv 2\pi f_0/f_s$

Difference equation of the digital second-order band-stop filter with coefficient formulas

Functional Block Diagram

The hum eliminator is simply a cascaded network of $N = 5$ band-stop filters, where the series network is similar to the cascaded networks of Chapters 12 through 15. Since the frequency components that need to be filtered are multiples of the fundamental frequency $f_0 = 60$ Hz, the kth band-stop section's center frequency and quality factor are determined by the following:

$$\begin{aligned}\theta_k &= k\theta_0 \\ &= 2\pi k f_0/f_s\end{aligned} \quad (18\text{-}2a)$$

$$\begin{aligned}Q_k &= kQ_0 \\ &= kf_0/B\end{aligned} \quad (18\text{-}2b)$$

where B is the bandwidth in Hz. Combining Equations (18-1b), (18-2a), and (18-2b), results in a new description of the coefficients, as shown in Equation (18-3b). An interesting feature of this result is that the β and α are identical between sections, whereas γ is different for each section.

Figure 18-5 shows the total network response of the cascaded band-stop filters based on Equations (18-3) and the network of *Figure 18-4*, where $B = 10$ Hz.

60 Hz Hum Eliminator

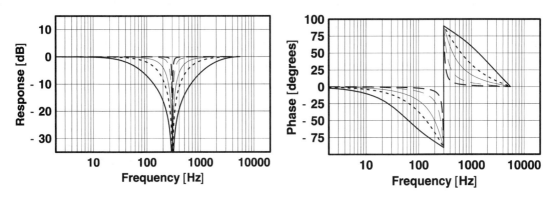

Figure 18-3. Gain response and phase plots for IIR band-stop filter with $f_s = 44100$ and $f_0 = 300$: solid line, Q = 0.2; dotted line, Q = 0.5; thin solid line, Q = 1; thin dashed line, Q = 3; dashed line, Q = 10

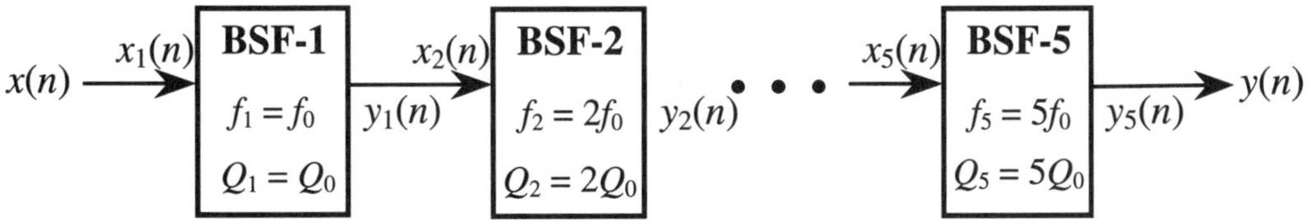

$$x(n) \rightarrow \boxed{\begin{array}{c} \text{BSF-1} \\ f_1 = f_0 \\ Q_1 = Q_0 \end{array}} \xrightarrow{x_2(n)} \boxed{\begin{array}{c} \text{BSF-2} \\ f_2 = 2f_0 \\ Q_2 = 2Q_0 \end{array}} \xrightarrow{y_2(n)} \cdots \xrightarrow{x_5(n)} \boxed{\begin{array}{c} \text{BSF-5} \\ f_5 = 5f_0 \\ Q_5 = 5Q_0 \end{array}} \xrightarrow{y_5(n)} y(n)$$

Figure 18-4. Block diagram of hum eliminator: $f_0 = 60$ Hz

$$y_k(n) = 2\{\alpha\, x_k(n) - \gamma_k\, x_k(n-1) + \alpha\, x_k(n-2) + \gamma_k\, y_k(n-1) - \beta\, y_k(n-2)\} \quad (18\text{-}3a)$$

$$\beta = \frac{1}{2}\frac{1-\tan(\pi B/f_s)}{1+\tan(\pi B/f_s)} \qquad \gamma_k = \left(\tfrac{1}{2}+\beta\right)\cos k\theta_0 \qquad \alpha = \left(\tfrac{1}{2}+\beta\right)/2 \quad (18\text{-}3b)$$

$$\text{where,} \quad \theta_0 \equiv 2\pi_0 f/f_s \qquad f_0 = 60\,Hz$$

Difference equation of the IIR second-order band-stop filter with coefficient formulas, specifically for suppressing harmonic frequencies of the fundamental frequency f_0. Note B is bandwidth in Hz

DSP Filters

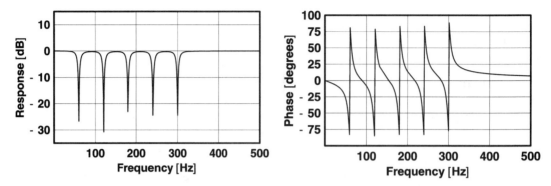

Figure 18-5. Total network response of the cascaded bandstop filters, based on Equations (18-3) and the network of Figure 18-4. Note that B = 10 Hz

Coefficient calculation section

The coefficient calculation section can be implemented by applying the gain values read from the front panel sliders to the equations derived in chapter 10. There are several potential variables (as shown by the equations in Chapter 10) that affect the band-stop filter operation, including sample rate, center frequency, Q factor, and gain. In the implementation described below, we have chosen to apply a constant-Q (4.2426) with fixed center frequencies (one-third octaves). The variables in our design are the filter gain and sample rate.

Filter section

Audio signals are by nature analog and must be converted to a numerical representation by an A/D converter in order to be filtered in the digital domain. Therefore, the audio signal provided to the filter section must first pass through an audio quality A/D converter. These are typically 16- to 24-bit converters. Filtering is then applied in the digital domain. Unlike bandpass filters, band-stop filters do not exhibit the side effect of attenuating the stop band to a theoretical negative infinity. Instead, the side bands are relatively unaffected, and outside of the frequency band of interest all others remain flat. This

60 Hz Hum Eliminator

characteristic allows us to implement a series of cascaded second order IIR band-stop filters. If we had chosen bandpass filters, as in the second implementation, we would have been required to implement them in parallel. The filter output is represented in the digital domain and must be converted to an analog signal in order for us to hear the results. The digital audio signal output from the filter section is then passed through an audio quality D/A converter.

Implementation

The implementation of the graphic equalizer is broken into two sections — coefficient calculation and filtering. The hum eliminator implementation is first described using flow diagrams followed by detailed descriptions of the various software modules. This implementation is designed to meet the requirements as outlined in the Design Requirements section of this chapter.

Flow diagram descriptions

The flow diagram in *Figure 18-6* represents the routine *calc_coef_for_new_gain(gain, i)*. This routine implements the coefficient calculation section. When a change in sample rate is made, the *calc_coef_for_new_gain(gain, i)* routine is called and passes both the band index *i* and the new sample rate. The new gain value, a dummy value for compatibility with the other projects, is first tested to make sure that it falls between the maximum and minimum values established by the system for the band. The new gain is saved and the process to calculate the coefficients begins, based on the formulas from Chapter 9. At this point, the coefficients beta, gamma, and alpha can be determined. Each of the variables and coefficients is stored for use by the filter corresponding to the index *i*.

The second operation (processing the samples) is shown in *Figures 18-6* and *18-8*. The flow diagram of *Figure 18-7* represents a routine

DSP Filters

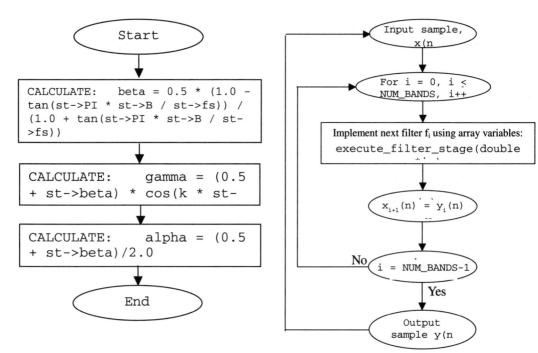

Figure 18-6. Coefficient calculation section
Routine: calc_coef_for_new_gain( ).

Figure 18-7. Filter section
Routine: execute_filter_block_in_place(double *in)

called *execute_filter_block_in_place( )*. This routine implements the filter section. *execute_filter_block_in_place( )* receives an input sample from the A/D converter, processes the sample through each of the five cascaded band-stop filters, and outputs the resulting sample to the D/A converter. The subroutine *execute_filter_stage( )* is called five times per sample. Each call implements one of the five cascaded band-stop filters.

The flow diagram in *Figure 18-8* represents a routine called *execute_filter_stage( )*. *execute_filter_stage( )* implements the band-stop filter. There is only one band-stop filter routine. It is called 31 times for every new sample by the *execute_filter_block_in_place( )* routine shown in *Figure 18-7*. The filter stage output is determined by the difference equation and calculated by the *execute_filter_stage( )* routine. All of the filter

60 Hz Hum Eliminator

states ($x_i(n-1)$, $x_i(n-2)$, $y_i(n-1)$, and $y_i(n-2)$) are updated and stored, as well as the filter output to be used for the next filter stage. The last filter in the section outputs the resulting sample to the D/A converter.

Software description

All filter-related information is stored in a two-dimensional array or matrix as shown in Table 18-1. The matrix is used to store variables required for processing each of the band-stop filters. Each column contains the variables associated with one of the 31 band-stop filters in the filter section and is referenced in the following routines by the value *i*. Each row represents a particular variable required for processing each of the band-stop filters.

Table 18-1. Bandstop Filter variable matrix: stages[i]->variable

i -> Variable	0	1	2	3	4
min_f0 – Minimum Bandstop Filter Center Frequency	0.1	0.1	0.1	0.1	0.1
f0 – Bandstop Filter Center Frequency	60	120	180	240	300
max_f0 – Max Bandstop Filter Center Frequency	22049	22049	22049	22049	22049
min_gain – Minimum Bandstop Filter Gain	0	0	0	0	0
Gain – Bandstop Filter Gain	1	1	1	1	1
max_gain – Maximum Bandstop Filter Gain	10	10	10	10	10
min_B – Minimum Bandstop Filter Bandwidth	0.1	0.1	0.1	0.1	0.1
B – Bandstop Filter Bandwidth	10	10	10	10	10
max_B – Maximum Bandstop Bandwidth	1000	1000	1000	1000	1000
x – Bandstop Filter Input Sample	0	0	0	0	0
x1 – Previous Bandstop Filter Input Sample	0	0	0	0	0
x2 – Previous, Previous Bandstop Filter Input Sample	0	0	0	0	0
y – Bandstop Filter Output Sample	0	0	0	0	0
y1 – Previous Bandstop Filter Output Sample	0	0	0	0	0
y2 – Previous, Previous Filter Output Sample	0	0	0	0	0
theta0 – $2*pi*f_0/f_s$	.00855	.01710	.02565	.03419	.04274
beta – Bandstop Filter Feedback Coefficient	0.499288	0.499288	0.499288	0.499288	0.499288
gamma – Bandstop Filter Feedback Coefficient	0.999252	0.999142	0.99896	0.998704	0.998375
alpha – Bandstop Filter Feed-Forward Coefficient	0.499644	0.499644	0.499644	0.499644	0.499644

DSP Filters

The following software routines implement the functional block diagram of *Figure 18-3*, as originally designed, into a DirectX application and follows the program flow diagrams described in *Figures 18-6* through *18-8*. The routines are all provided as software modules written in C++ covering each of the three major components.

Initialize the variable matrix

The implementation begins with a description of the initialization routine required to establish the variable matrix to a known stable state. Note that the table values are just an initial stable state. The values actually used for filter computation are calculated and filled into this table later (in step 2 – *Calculating the Coefficients*). Initialization is performed prior to activation of the hum eliminator. It is performed only once prior to activation or following any system reset condition. A line-by-line analysis of the initialization routine is provided below.

```
1   C60_Hz_Hum_Eliminator::C60_Hz_Hum_Eliminator()
2   {
3       int i;
4       min_beta = -0.4999;
5       max_beta = 0.4999;
6       stages = new (CCookFilterStage * [NUM_BANDS]);
7       for (i = 0; i < NUM_BANDS; i++)
8       {
9           stages[i] = new CBandstopFilterStage;
10          stages[i]->min_f0 = 0.1;
11          stages[i]->min_gain = -20.;
12          stages[i]->max_gain = 20.0;
13          stages[i]->min_Q = 0.1;
14          stages[i]->max_Q = 20.0;
15          stages[i]->gain = 0.0;
16          stages[i]->B = 10;
17          stages[i]->f0 = 60;
18          stages[i]->max_f0 = 22049.0;
19          calc_coeff_for_new_f0 (10000.0,i);
20      }
21      num_filter_stages = NUM_BANDS;
22      enabled = 1;
23  }
```

60 Hz Hum Eliminator

- Lines 1, 2, and 20 define the boundaries of this routine. Everything between these lines constitutes the initialization routine.

- Lines 3 through 6 define and initialize processing parameters as follows:

    Line 3 defines the variable i as an integer that is used to reference variables in a particular column in the matrix shown in Table 1. There are five frequency bands and thus i will operate from 0 to 4.

    Lines 4 and 5 establish the minimum and maximum beta coefficient values. Each time beta is calculated it will be compared to these values and forced to reside between them.

    Line 6 calls a routine that allocates the pointers used for the variable matrix.

- Lines 7 through 20 establish a loop that will be performed NUM_BANDS times. In this example, NUM_BANDS is defined in a separate file header and equals 5. This loop will initialize the variable matrix to known stable states as shown in Table 1. Each time the loop is implemented, *i* is incremented to reference a new column in the matrix to be initialized.

    Line 9 implements a constructor that establishes the matrix variables and associates them with the rows of the matrix.

    Lines 10 through 18 initialize each of the respective variables to the values indicated.

    Line 19 calls the routine calc_coeff_for_new_f0 (10000.0,i). This routine initializes the coefficients to a known stable value. This routine is described in more detail in step 2 *(Calculating the Coefficients)* below.

- Line 21 establishes num_filter_stages equal to NUM_BANDS, which in this case is 5.

- Line 22 indicates that initialization has been completed and the hum eliminator can now be enabled.

    The following routine is the constructor and initializes the filter state variables.

DSP Filters

```
1  CBandstopFilterStage::CBandstopFilterStage()
2  {
3      x = 0;
4      y = 0;
5      x1 = 0;
6      x2 = 0;
7      y1 = 0;
8      y2 = 0;
9  }
```

Following the completion of the initialization routine, the matrix fills out to the values shown in Table 1-18.

Calculating the coefficients

When the updated gain value has been received for a particular band, the routine is called to calculate the filter coefficients. Note from the equations in Chapter 9 that the coefficients are a function of the sample rate, center frequency, Q factor, and gain. Only gain and sample rate are adjustable by the user. The center frequencies are automatically calculated in line 19 of the initialization routine *C60_Hz_Hum_Eliminator( )* and are not adjustable by the user. The Q factor is replaced with a constant bandwidth B, for the hum eliminator project.

Although a user interface to control sample rate is not usually provided, automatic sample rate changes are common and when the system detects a new sample rate, the routine *calc_coef_for_new_fs (double dv)* should automatically calculate new coefficients based on sample rate changes. Note that this routine modifies variables across every column of the matrix. This means it affects each of the 31 filters implemented in the filter section. A description of this routine is provided here:

```
1  void C60_Hz_Hum_Eliminator::calc_coeff_for_new_fs (double dv)
2  {
3      int i;
4      fs = dv;
5      for (i = 0; i< NUM_BANDS; i++)
```

```
 6        {
 7                stages[i]->max_f0 = fs/2.0 - 1.0;
 8                calc_coeff_for_new_f0 (0.0, i);
 9        }
10 }
```

- Lines 1, 2, and 10 define the boundaries of this routine. The routine receives one variable; new sample rate. The new sample rate is denoted by the variable name *dv*.

- Line 3 defines the variable *i* as an integer used to reference variables in a particular column in the matrix shown in Table 1. There are five frequency bands and thus *i* will operate from 0 to 4. Each iteration is referred to as a stage in the graphic equalizer. Each stage implements one band-stop filter and has a dedicated column in the matrix allocated for its use.

- Line 4 saves the new sample rate *dv* as the variable *fs*.

- Line 5 begins a loop that will process five iterations, one for each filter section.

- Lines 6-9 perform the function of modifying matrix variables.

    *Line 7 calculates a new maximum center frequency (*max_f0*) value based on Nyquist's theorom. The maximum center frequency should not exceed one-half of the sample frequency.*

    Line 8 calls the routine calc_coeff_for_new_f0 (0.0,i), *which then updates additional variables in the matrix related to sample rate.*

The routine *calc_coeff_for_new_f0 (double dv ,int i)* is shown below. This routine receives an integer representing the column associated with the particular filter being operated on. A description of this routine is provided here:

DSP Filters

```
1 void C60_Hz_Hum_Eliminator::calc_coeff_for_new_f0 (double dv, int i)
2 {
3      CCookFilterStage *st = stages[i];
4      st->theta0 = TWO_PI * st->f0/fs;
5      calc_coeff_for_new_gain (st->gain, i);
6 }
```

- Lines 1, 2, and 6 define the boundaries of this routine.
- Line 3 establishes *st* as a pointer to the variables in column *i* of the variables matrix.
- Line 4 calculates a new theta based on the center frequency and the updated sample rate for the respective column.
- Line 5 calls the routine *calc_coeff_for_new_gain (st->gain, i)*. The gain is passed to this routine along with *i*. This routine calculates the new coefficients beta, gamma, and alpha, giving the updated variables based on center frequency and sample rate changes.

The routine *calc_coeff_for_new_gain(double dv, int i)* calculates the filter coefficients by implementing the equations defined in Chapter 9. A description of this routine is provided below:

```
1 void C60_Hz_Hum_Eliminator::calc_coeff_for_new_gain (double dv, int i,k)
2 {
3      CCookFilterStage *st = stages[i];
4      if (dv > st->max_gain) dv = st->max_gain;
5      else if (dv < st->min_gain) dv = st->min_gain;
6      st->gain = dv;
7      k = i + 1;
8      st->beta = 0.5 * (1.0 - tan(st->PI * st->B / st->fs)) /
9               (1.0 + tan(st->PI * st->B / st->fs));
10     st->gamma = (0.5 + st->β) * cos(k * st->θ0);
11     st->alpha = (0.5 + st->β)/2.0;
12 }
```

60 Hz Hum Eliminator

- Lines 1, 2, and 12 define the boundaries for this routine.
- Line 3 establishes *st* as a pointer to the variables in column i.
- Line 4 tests the new *gain*, verifying that it does not exceed the maximum gain boundaries *(max_gain)*.
- Line 5 tests the new *gain*, verifying that it is not lower than the minimum gain boundaries *(min_gain)*.
- Line 6 saves the new gain in column i of the matrix.
- Line 7 sets index k equal to i + 1.
- Lines 8 and 9 calculate a new β value.
- Line 10 calculates a new β value.
- Line 11 calculates a new α value.

Implement the band-stop filter

The filter section as shown in *Figure 18-6* is implemented in the routine *execute_filter_block_in_place(double *in)*. A description of this routine is provided below:

```
1 void C60_Hz_Hum_Eliminator::execute_filter_block_in_place(double *in)
2 {
3       int i;
4       double input = *in;
5       stages[0]->x = input;
6       stages[0]->execute_filter_stage();
7       for (i= 1; i< NUM_BANDS; i++)
8       {
9               stages[i]->x = stages[i-1]->y;
10              stages[i]->execute_filter_stage();
11      }
12      *in = stages[NUM_BANDS-1]->y;
13 }
```

DSP Filters

- Lines 1, 2, and 13 define the boundaries of this routine.
- Line 3 defines the variable *i* as an integer that is used to reference variables in a particular column in the matrix shown in Table 1. There are 31 frequency bands and thus *i* will operate from 0 to 30. Each iteration is referred to as a stage in the graphic equalizer. Each stage implements one band-stop filter and has a dedicated column in the matrix allocated for its use.
- Line 4 defines a variable input initialized to the pointer value *in. *in is a pointer to the next input sample.
- Line 5 stores the input sample value in the first filter's variable x in the first column of the matrix.
- Line 6 calls the routine *execute_filter_stage( )* which processes the first filter.
- Lines 7-11 contain a loop which processes the last 30 filters in cascaded fashion.
- Line 9 moves the output of the previous filter to the input of the present filter.
- Line 10 calls the routine *execute_filter_stage( )* to operate using the present filter stage variables.
- Line 12 replaces the input sample, pointed to by *in, with the output value of the last filter stage. The output value of the last filter is the output of the graphic equalizer. *in is thus pointing to the graphic equalizer's output sample. It will be passed to the D/A converter.

The *execute_filter_stage( )* implements the second-order IIR band-stop filter as shown in the flow diagram in *Figure 18-7*. The coefficients and state variables are stored in the respective columns of the matrix for each stage. The variables utilized in this routine are listed here:

60 Hz Hum Eliminator

```
x                     - input sample
x1, x2                - past input samples
y                     - output sample
y1, y2                - past output samples
α, γ, β       - filter coefficients
```

This routine implements the difference equation shown in *Figure 18-8*. The following describes this routine:

```
1 void CBandstopFilterStage::execute_filter_stage()
2 {
3     y = 2 * (α * x - γ * x1 + α * x2 + γ * y1 - β * y2);
4     x2 = x1;
5     x1 = x;
6     y2 = y1;
7     y1 = y;
8 }
```

- Lines 1, 2, and 8 define the boundaries of this routine.
- Line 3 calculates the output value y for the corresponding filter.
- Lines 4 through 7 update the filter state variables (x1, x2, y1, and y2) for the corresponding filter.

As expressed in the equations derived in Chapter 9, the coefficients (beta, gamma, and alpha) are a function of the sample rate, center frequency, Q-factor (or bandwidth B), and the gain of the band-stop filter. Therefore, the first time *execute_filter_stage()* routine is called for a given sample, the coefficients (β_i, γ_i, and α_i) in column 0 are used to filter the input sample. The second time it is called, the coefficients in column 1 are used, etc. Each time this routine is called, it must update the past variable states ($x_i(n-1)$, $x_i(n-2)$, $y_i(n-1)$, and $y_i(n-2)$) in the respective column of the matrix as well as the output $y_i(n)$. Each band-stop filter cascaded in the filter section copies its output sample $y_i(n)$ to the input of the following filter $x_{i+1}(n)$ as shown in the flow diagram. The last filter in the section outputs the resulting sample to the D/A converter.

DSP Filters

During shutdown of the hum eliminator, a destructor routine can be used to release the memory making it available to the system for other applications. The routine ~C60_Hz_Hum_Eliminator( ) as shown below can be used for this purpose.

```
1  C60_Hz_Hum_Eliminator::~C60_Hz_Hum_Eliminator()
2  {
3      int i;
4      if (stages)
5      {
6          for (i = 0; i< NUM_BANDS; i++)
7          {
8              delete stages[i];
9          }
10         delete stages;
11         stages = 0;
12     }
13 }
```

19

31-Band Graphic EQ-I

Introduction

This chapter describes the first of two graphic equalizer implementations. Graphic equalizers can be implemented using either the peaking filter or the bandpass filter. Both implementations are common and have unique characteristics. This chapter is a guide providing step-by-step instructions for implementing a 31-band graphic equalizer using the peaking filter described in Chapter 10. The following chapter describes the same 31-band graphic equalizer implemented using the bandpass filter developed in Chapter 8.

Graphic equalizers are processors that modify the frequency response of an audio signal by dividing its audible frequency spectrum (~20 Hz to ~20 kHz) into several frequency bands and alter the frequency response of each band independently by user controls. Graphic

DSP Filters

equalizers apply a set of bandpass or peaking filters to modify the frequency response of the incoming audio signal. These filters maintain constant center frequencies and Q factors but support variable gains. Sliders on the front panel of the equalizer are made available to control the filter gain as shown in *Figure 19-1*. The position of the individual sliders determines the amount of gain or cut applied to the filter operating on the corresponding band and when the sliders are viewed side-by-side, they graphically resemble the frequency response applied to the input audio signal. Therein lies the source of the device name, graphic equalizer.

Graphic equalizers serve many purposes and their function typically depends upon the environment in which they are applied. In general, graphic equalizers are used to compensate for limitations in an audio system, room acoustics, or to accommodate listening preferences. For example, in a home theater system a graphic equalizer may be employed to compensate for limitations in the audio system (electronics, speakers) or the listening room. In an automobile, a graphic equalizer can be designed to boost signals masked by engine or road noise and to compensate for other audio system limitations.

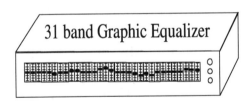

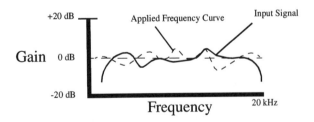

Figure 19-1. Graphic equalizer with corresponding frequency response

In live performances, such as in an auditorium, music hall, church, or gymnasium an equalizer can assist in suppressing feedback caused by signal peaks at particular frequencies related to room size and dimensions. A graphic equalizer may allow an audio engineer to neutralize a room's natural sound coloration or frequency response providing consistent performances regardless of the room's natural acoustical effects. It can also remove annoying 60 Hz hum associated with ground loops and electromagnetic interference from power sources. In professional settings such as a recording studio, graphic equalizers are used to sculpt sound in a more artistic manner. Adding effects to a recording or live performance like boosting the bass frequencies in particular instruments or even to accentuate a vocalist that may otherwise be masked by other instruments, can be performed with graphic equalizers. This type of application, however, is more commonly served with parametric equalizers, such as the one described in Chapter 21. It is likely that somewhere we have all experienced the benefits of graphic equalizers.

Although the uses of graphic equalizers have not dramatically changed since their inception, the implementation has evolved from conventional passive and active analog designs to the present barrage of digital implementations.

Design Requirements

The goal of this design is to develop a digital 31-band graphic equalizer meeting the following specifications:

Specifications:
Channels: 2 (stereo)
Bands: (31) with center frequencies at 1/3 octave intervals
Gain: +/- 20 dB
Q factor: $3\sqrt{2}$
Sample Rate: Variable

DSP Filters

This design will support stereo (two channels) audio signals with independent processing for each channel having 31 bands at one-third octave intervals. Each filter will use a constant-Q factor for predictability and minimal interaction between bands. The boost/cut range is ±20 dB with controllable step size. This graphic equalizer will be implemented using cascaded peaking filters as developed from equations described in Chapter 10.

Peaking Filter Overview

A quick review of the peaking filter is helpful at this point. Equations (19-1a), (19-1b), and (19-1c) are conveniently provided here for implementing a peaking filter along with the corresponding frequency and phase response curves.

Remember from Chapter 10 that a second-order IIR *peaking* filter can be implemented by summing the input x(n) with the output of a second-order bandpass filter scaled by µ-1, as shown in *Figure 19-2*. The bandpass output scale factor is chosen so that when µ = 1, the output is equal to the input, y(n) = x(n). The $_\mu$ coefficient depends on the peaking level g, as $\mu \equiv 10^{g/20}$, where g is the boost/cut gain in dB.

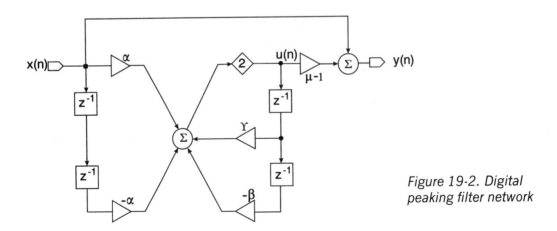

Figure 19-2. Digital peaking filter network

31-Band Graphic EQ-I

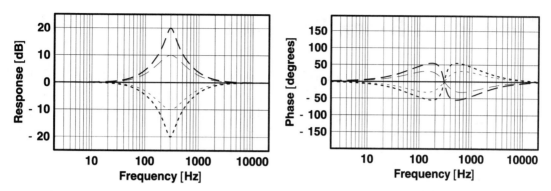

Figure 19-3. Gain and phase of digital peaking filter with Q = 1, f_0 = 300, and f_s = 44100: dotted line, g = -20 dB; thin dotted line, g = -10 dB; solid line, g = 0 dB; thin dashed line, g = +10 dB; dashed line, g = +20 dB

The frequency and phase response of the peaking filter is displayed in *Figure 19-3*. Note that since the network is second order, the phase excursion will not exceed ±90°. The maximum and minimum phase value is controlled by the shelving filter's boost/cut gain factor g.

Equations (19-1a), (19-1b), and (19-1c) are extracted from Chapter 10 and displayed here for convenience. The *center frequency* (f_0), *sample rate* (f_s), *gain* (g), and *Q factor* (Q) each will influence the calculation of the coefficients α, β, and γ in Equation (19-1c). α, β, and γ coefficients are then utilized to resolve the difference equation in Equation (19-1a) and calculate the output y(n) from Equation (19-1b).

$$u(n) = 2\{\alpha\,[x(n) - x(n-2)] + \gamma\,u(n-1) - \beta\,u(n-2)\} \qquad (19\text{-}1a)$$

$$y(n) = x(n) + (\mu - 1)\,u(n) \qquad (19\text{-}1b)$$

$$\beta = \frac{1}{2}\left(\frac{1 - \left(\frac{4}{1+\mu}\right)\tan\frac{\theta_0}{2Q}}{1 + \left(\frac{4}{1+\mu}\right)\tan\frac{\theta_0}{2Q}}\right) \qquad \gamma = \left(\tfrac{1}{2} + \beta\right)\cos\theta_0 \qquad \alpha = \left(\tfrac{1}{2} - \beta\right)/2 \qquad (19\text{-}1c)$$

$$\text{where,}\quad \theta \equiv 2\pi f / f_s \qquad \theta_0 \equiv 2\pi f_0 / f_s$$

Difference equation of the digital second-order peaking filter with coefficient formulas

DSP Filters

Functional Block Diagram

The graphic equalizer consists of three components as shown in the system level block diagram of *Figure 19-4*. The slider section is controlled by the user and is utilized to adjust the gain (g_i) applied to each frequency band's coefficient calculation block. The function of the coefficient calculation blocks is to calculate the 31 sets of coefficients (α, β, and γ) applied to each of the 31 filters. The filter section utilizes the α, β, and γ coefficients calculated for each frequency band to filter the input audio signal. The following is an overview of each section of the graphic equalizer.

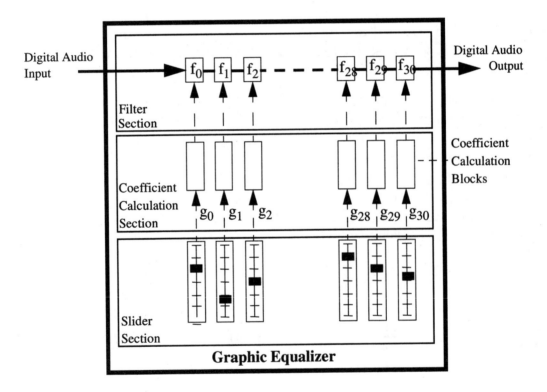

Figure 19-4. Graphic equalizer system level block diagram

Slider section

The control section of the graphic equalizer can be implemented in either hardware or in software. Regardless of the type of design required (hardware or software), the control section must provide a numerical representation of the gain controls to the coefficient calculation section.

Coefficient calculation section

The coefficient calculation section is implemented by applying the gain values read from the front panel sliders to the Equation (19-1c). There are several potential variables as shown by these equations that affect the peaking filter operation including sample rate (f_s), center frequency (f_o), Q factor, and gain (g). Typically, graphic equalizers provide the user with slider controls for gain, however the Q factor is made constant as well as the sample rate and center frequencies.

Filter section

Audio signals are by nature analog and must be converted to a numerical representation (samples) by an A/D converter in order to be filtered in the digital domain. The filtering section applies the difference equation, Equation (19-1a), and scales the output by Equation (19-1b). Unlike bandpass filters, peaking filters do not exhibit the side effect of attenuating the stop band to a theoretical negative infinity. Instead, peaking filters provide for relatively flat frequency response outside of the band of interest and the side bands are relatively unaffected. This characteristic allows us to implement cascaded second-order IIR peaking filters. If we had chosen bandpass filters — as are used in Chapter 20 — we would have been required to implement the filters in parallel. The filter output is represented in the digital domain and must be converted to an analog signal in order for us to hear the results. The digital audio signal output from the filter section is then passed through an audio quality D/A converter.

Flow Diagram Descriptions

The three sections as shown in *Figure 19-4* of the graphic equalizer are the slider section, coefficient calculation section, and filter section. The software implementation for the coefficient calculation section and filter section is first described using flow diagrams followed by the software routines and corresponding descriptions. This implementation is designed to meet the requirements as outlined in the design requirements section of this chapter.

As shown in *Figure 19-5*, there are four potential system level variables that affect the coefficients required to implement the graphic equalizer. The sample rate (f_s), center frequency (f_o), Q factor (Q), and gain factor (g) each affect the resulting coefficients used to implement the graphic equalizer. The coefficients (β, γ, and α) change dynamically with changes to any of these variables. Note that the Q factor is fixed in this implementation and the center frequencies are fixed at one-third octaves.

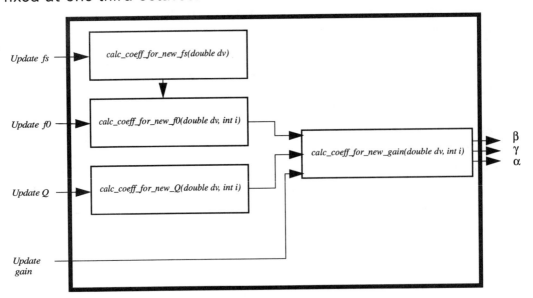

Figure 19-5. Coefficient calculation block diagram

31-Band Graphic EQ-I

There are two asynchronous operations taking place to implement the graphic equalizer. The first operation involves processing the audio samples in the filter section.

The second operation is reading the gains and updating the coefficients. The first operation is performed continuously at a fixed rate while the second operation only occurs when a change is made to one of the variables. Therefore, these operations are asynchronous.

Slider section

The slider section of the graphic equalizer can be implemented in either hardware or in software. Regardless of the type of design required (hardware or software), the slider section must provide a numerical representation of the slider positions.

Coefficient calculation section

Figures 19-5 through 19-8 display flow diagrams utilized for implementing the coefficient calculation section. As shown in *Figure 19-5*, there are four system level variables that effect the coefficients ß, γ, and α required to implement the graphic equalizer. The sample rate (f_s), center frequency (f_0), Q factor (Q), and gain factor (g) each affect the resulting coefficients used to implement the graphic equalizer. The coefficients change dynamically with changes to any of these variables.

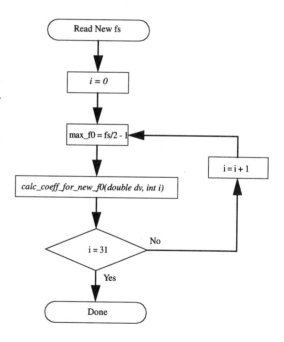

Figure 19-6. Flow diagram for updating fs Routine: calc_coeff_for_new_fs(double dv)

DSP Filters

Updating the sample rate

When the sample rate of the input data changes, the calc_coeff_for_new_fs(double dv) routine is called as shown by the flow diagram in Figure 19-6. This routine receives only one variable dv that contains the new sample rate (f_s) in Hertz. The sample rate change affects all filters and, therefore, no index i is passed to this operation. The change will be applied to each filter from this routine. The calc_coeff_for_new_fs (double dv) routine implements a loop where the index i is incremented from 0 to 30. Each iteration of the loop establishes an updated maximum center frequency based on the new sample rate and the Nyquist theorem for each filter band. calc_coeff_for_ new_f₀(double dv, int i) is called each iteration of the loop and updates the corresponding normalized center frequency (θ) and ultimately the respective coefficients. This routine is shown in Figure 19-7 and described below.

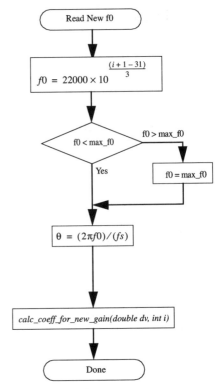

Figure 19-7. Flow diagram for updating f_0 Routine: calc_coeff_for_new_f_0(double dv, int i)

Updating the center frequency

The calc_coeff_for_new_f_0 (double dv, int i) routine shown in Figure 19-7 is called during initialization to establish the center frequencies for each filter. It is also activated when a sample rate change occurs. Two variables are passed to this routine although the center frequency

31-Band Graphic EQ-I

variable *dv* is actually a dummy variable required to conform to the format specification in the DirectX implementation. It is ignored and has no function in this routine since the center frequencies are fixed at one-third octaves. The index *i* variable indicates which of the 31 filters is to be processed with updated data. The fixed center frequency (f_0) that is calculated in this routine is tested to verify that it does not exceed the Nyquist frequency (max_f_0). The normalized center frequency (θ) is calculated as a function of the center frequency and the sample rate, and then the routine calc_coeff_for_new_gain (double dv, int i) is called to calculate the coefficients for the associated filter based on the new θ value.

Updating the gain

The flow diagram in *Figure 19-8* represents the routine calc_coef_for_new_gain(double dv, int i). This routine implements the coefficient calculation blocks. There are two variables passed to this routine — the gain (*dv*) and the index *i*. The index *i* indicates to which band the calc_coef_for_new_gain(double dv, int i) routine is to be applied. The variable *dv* represents the new gain value for that particular band. The new gain value is first tested to make sure that it falls between the maximum (*max_gain*) and minimum (*min_gain*)values established by the system for the band. If the new gain value

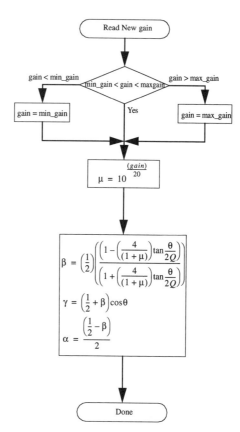

Figure 19-8. Flow diagram for updating gain Routine: calc_coeff_for_new_gain(double dv, int i)

233

DSP Filters

exceeds the maximum gain value it is set equal to the maximum value. If it falls below the minimum gain value it is set equal to the minimum value. The new gain is then saved and the process to calculate the coefficients begins based on Equation (19-1c). The μ coefficient is calculated as a function of the gain. Each of the variables and coefficients is stored for use by the filter corresponding to the index i.

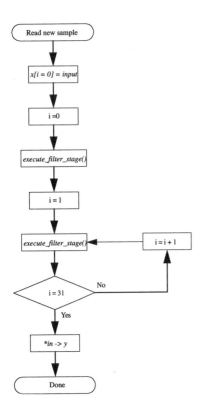

Figure 19-9. Flow diagram for processing input sample
Routine: execute_filter_block_in_place(double *in)

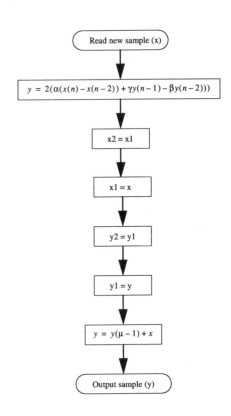

Figure 19-10. Flow diagram for peaking filter implementation
Routine: execute_filter_stage()

Filter section

The filtering process is shown in *Figures 19-9* and *19-10*. The flow diagram of *Figure 19-9* represents a routine called *execute_filter_block_in_place(double *in)*. This routine implements the filter section. *execute_filter_block_in_place(double *in)* receives an input sample pointed to by **in*, processes the sample through each of the 31 cascaded peaking filters, and outputs the resulting sample. The subroutine *execute_filter_stage( )* is called 31 times per sample. Each time it is called it implements one of the 31 cascaded peaking filters. *Figure 19-10* shows the *execute_filter_stage()* flow diagram. This routine implements the difference equation in Equation (19-1a) and scales the output per Equation (19-1b).

Software Description

All filter-related information is stored in a two-dimensional matrix as shown in Table 1-19. The matrix is used to store variables required for processing each of the peaking filters. Each column contains the variables associated with one of the 31 peaking filters and is indexed in the following routines by the index variable *i*. Each row represents a particular variable that is required for processing each of the peaking filters.

The following software routines implement the functional block diagram of *Figure 19-4* as originally implemented in a DirectX application and follows the program flow diagrams described in *Figures 19-5* through *19-10*. The routines are all provided as software modules written in C++.

Initialize the variable matrix

The implementation begins with a description of the initialization routine required to initialize the matrix to a known stable state. Initialization is performed only once prior to activation or following any sys-

DSP Filters

Table 1-19: Peaking Filter Variable Matrix: stages[i]->variable

Variable	i ->	0	1	2	....	....	28	29	30
min_f0 - Minimum Peaking Filter Center Frequency		0.1	0.1	0.1	....	....	0.1	0.1	0.1
f0 - Peaking Filter Center Frequency		21.5	27	34	....	....	13859	17461	22000
max_f0 - Maximum Peaking Filter Center Frequency		22049	22049	22049	....	....	22049	22049	22049
min_gain - Minimum Peaking Filter gain		-20	-20	-20	....	....	-20	-20	-20
Gain - Peaking Filter gain		0	0	0	....	....	0	0	0
max_gain - Maximum Peaking Filter gain		20	20	20	....	....	20	20	20
min_Q - Minimum Peaking Filter Q factor		0.1	0.1	0.1	....	....	0.1	0.1	0.1
Q - Peaking Filter Q factor		4.2	4.2	4.2	....	....	4.2	4.2	4.2
max_Q - Maximum Peaking Q factor		20	20	20	....	....	20	20	20
x - Peaking Filter Input Sample		0	0	0	....	....	0	0	0
x1 - Previous Peaking Filter Input Sample		0	0	0	....	....	0	0	0
x2 - Previous, Previous Peaking Filter Input Sample		0	0	0	....	....	0	0	0
y - Peaking Filter Output Sample		0	0	0	....	....	0	0	0
y1 - Previous Peaking Filter Output Sample		0	0	0	....	....	0	0	0
y2 - **Previous,** Previous Peaking Filter Output Sample		0	0	0	....	....	0	0	0
mu - $10^{G/20}$		1	1	1	....	....	1	1	1
theta0 - $2*pi*f_0/f_s$		.00306	.00385	.00484	....	....	1.9746	2.4878	3.1345
beta - Peaking Filter feedback coefficient		.4996	.4995	.4994	....	....	.3084	.2681	.2209
gamma - Peaking Filter feedback coefficient		-.6277	-.6277	-.6277	....	....	-.6277	-.6277	-.6277
alpha - Peaking Filter feed-forward coefficient		-.6277	-.6277	-.6277	....	....	-.6277	-.6277	-.6277

tem reset condition. A line-by-line analysis of the initialization routine is provided below:

```
1  CFixed_31_Band_Peaking::CFixed_31_Band_Peaking()
2  {
3      int i;
4      min_beta = -0.4999;
5      max_beta = 0.4999;
6      stages = new (CCookFilterStage * [NUM_BANDS]);
7      for (i = 0; i < NUM_BANDS; i++)
8      {
9          stages[i] = new CPeakingFilterStage;
10         stages[i]->min_f0 = 0.1;
11         stages[i]->min_gain = -20.;
12         stages[i]->max_gain = 20.0;
13         stages[i]->min_Q = 0.1;
14         stages[i]->max_Q = 20.0;
15         stages[i]->gain = 0.0;
16         stages[i]->Q = sqrt(2.0) * 3.0;
```

31-Band Graphic EQ-I

```
17          stages[i]->f0 = 22000.0 * pow(2.0,((double)i+1.0-
NUM_BANDS)/3.0);
18          stages[i]->max_f0 = 22049.0;
19          calc_coeff_for_new_f0 (10000.0,i);
20      }
21      num_filter_stages = NUM_BANDS;
22      enabled = 1;
23  }
```

- Lines 1, 2, and 23 define the boundaries of this routine. Everything between these lines constitutes the initialization routine.

- Lines 3 through 6 define and initialize the processing parameters as follows:

    Line 3 defines the index variable i as an integer that is used to reference variables in a particular column in the matrix shown in Table 1-19. There are 31 frequency bands and thus i can contain any of the values from 0 to 30.

    Lines 4 and 5 establish the minimum and maximum b coefficient values. Each time b is calculated it will be compared to these values and forced to reside between them.

    Line 6 calls a routine that allocates the pointers to be used for the variable matrix.

- Lines 7 through 20 establish a loop that will be performed NUM_BANDS times. NUM_BANDS represents the number of frequency bands and is defined in a separate file header to equal 31. This loop will initialize the matrix to a known stable state as shown in Table 1-19. Each time the loop is implemented index *i* is incremented to reference a new column in the matrix to be initialized.

    Line 9 invokes a constructor that saves memory in the system for the matrix variables and associates the variables with the rows of the matrix.

    Lines 10 through 18 initialize each of the respective variables to the values indicated.

    Line 19 calls the routine calc_coeff_for_new_f0 (10000.0,i). This routine utilizes the previously initialized variables to initialize the center frequencies and coefficients to a known stable value. This routine is described in more detail in a subsequent section below. Note that the value 1000.00 is a dummy variable and has no significance in the initialization process.

DSP Filters

- Line 21 establishes *num_filter_stages* equal to NUM_BANDS, which in this case is 31. *num_filter_stages* is not relevant to the operation of the graphic equalizer but is required overhead in the DirectX implementation.

- Line 22 indicates that initialization has been completed and the equalizer can now be enabled.

The following routine is the constructor that initializes the filter state variables:

```
1  CPeakingFilterStage::CPeakingFilterStage()
2  {
3      x = 0;
4      y = 0;
5      x1 = 0;
6      x2 = 0;
7      y1 = 0;
8      y2 = 0;
9  }
```

Following the completion of the initialization routine, the matrix reflects the values shown in Table 1-19.

Coefficient calculation section

When the updated gain value has been received for a particular band the *calc_coeff_for_new_gain(double dv, int i)* routine is called to calculate the filter coefficients.

Typically the sample rate is a fixed value in a digital graphic equalizer. However, this is not always the case. In a design where the sample rate can change the routine *calc_coef_for_new_fs (double dv)* should be called to calculate new coefficients based on the sample rate changes. Note that this routine modifies variables across every column of the matrix. This means it affects each of the 31 filters implemented in the filter section. A description of this routine is provided here:

31-Band Graphic EQ-I

```
1  void CFixed_31_Band_Peaking::calc_coeff_for_new_f_s (double dv)
2  {
3      int i;
4      fs = dv;
5      for (i = 0; i< NUM_BANDS; i++)
6      {
7          stages[i]->max_f_0 = f_s/2.0 - 1.0;
8          calc_coeff_for_new_f_0 (0.0, i);
9      }
10 }
```

- Lines 1, 2, and 10 define the boundaries of this routine. The routine receives one variable: new sample rate. The new sample rate is denoted by the variable name *dv*.

- Line 3 defines the index variable *i* as an integer that is used to reference variables in a particular column in the matrix shown in Table 1-19. There are 31 frequency bands and thus *i* will operate from 0 to 30. Each iteration is referred to as a stage in the graphic equalizer. Each stage implements one peaking filter and has a dedicated column in the matrix allocated for its use.

- Line 4 saves the new sample rate *dv* as the variable f_s.

- Line 5 begins a loop consisting of lines 6-9 that will operate across each of he 31 bands.

- Lines 6 through 9 perform the function of modifying matrix variables.

    Line 7 calculates a new maximum center frequency (max_f_0) value based on Nyquist's theorem which states that the maximum center frequency should not exceed fs/2.

    Line 8 calls the routine calc_coeff_for_new_f_0 (0.0,i) that updates additional variables in the matrix related to sample rate as shown below. Note that the value 0.0 is a dummy variable and has no significance.

DSP Filters

The routine *calc_coeff_for_new_f$_0$* (double dv ,int i) is shown below. This routine can be called directly whenever the center frequency of a band is changed or when the sample rate is changed. This routine receives a dummy value *dv* and an index *i* representing the column associated with the particular filter being operated on. The dummy value is a place saver in the DirectX implementation although not used. A description of this routine is provided below:

```
1  void CFixed_31_Band_Peaking::calc_coeff_for_new_f0 (double dv, int i)
2  {
3       CCookFilterStage *st = stages[i];
4       st->f0 = 22000.0 * pow(2.0,((double)i+1.0-NUM_BANDS)/3.0);
5       if (st->f0 > st->max_f0)
6       {
7            st->f0 = st->max_f0;
8       }
9       st->theta0 = TWO_PI * st->f0/fs;
10      calc_coeff_for_new_gain (st->gain, i);
11 }
```

- Lines 1, 2, and 11 define the boundaries of this routine.

- Line 3 establishes *st* as a pointer to the variables in column *i* of the variables matrix.

- Line 4 calculates a new center frequency for the peaking filter associated with the column being operated on.

- Lines 5 through 8 test the center frequency, verifying that it does not exceed the maximum center frequency boundary *(max_f$_0$)*. If f_0 > *max_f$_0$*, then f_0 = *max_f$_0$*.

- Line 9 calculates a new normalized center frequency (*q*) based on the new center frequency and sample rate for the respective column *i*.

- Line 10 calls the routine *calc_coeff_for_new_gain (st->gain, i)*. The gain is passed to this routine along with the index *i*. This routine calculates the new coefficients ß, γ, and α given the updated variables based on center frequency and sample rate changes.

31-Band Graphic EQ-I

The routine *calc_coeff_for_new_gain(double dv, int i)* can be called directly when the gain is changed by a slide adjustment for any of the channels or from either of the routines above. This routine calculates the filter coefficients by implementing the Equation (19-1c). A description of this routine is provided below:

```
1  void CFixed_31_Band_Peaking::calc_coeff_for_new_gain (double dv, int i)
2  {
3       CCookFilterStage *st = stages[i];
4       if (dv > st->max_gain) dv = st->max_gain;
5       else if (dv < st->min_gain) dv = st->min_gain;
6       st->gain = dv;
7       st->mu = pow(10.0,st->gain/20.0);
8       st->beta = 0.5 * (1.0 - (4.0/(1.0+st->mu))*tan(st->theta0/(2.0*st->Q))) /
9                        (1.0 + (4.0/(1.0+st->mu))*tan(st->theta0/(2.0*st->Q)));
10      st->gamma = (0.5 + st->beta) * cos(st->theta0);
11      st->alpha = (0.5 - st->beta)/2.0;
12 }
```

- Lines 1, 2, and 12 define the boundaries for this routine.
- Line 3 establishes *st* as a pointer to the variables in column i.
- Line 4 tests the new gain verifying that it does not exceed the maximum gain boundary *(max_gain)*.
- Line 5 tests the new gain verifying that it is not lower than the minimum gain boundary *(min_gain)*.
- Line 6 saves the new gain in column *i* of the matrix.
- Line 7 calculates a new μ value in column *i* of the matrix.
- Lines 8 and 9 calculate a new β value in column *i* of the matrix.
- Line 10 calculates a new γ value in column *i* of the matrix.
- Line 11 calculates a new α value in column *i* of the matrix.

DSP Filters

Peaking filter section

The filter section applies the coefficients to Equations (19-1a) and (19-1b) in the routine *execute_filter_block_in_place(double *in)*. Figure 19-9 displays the flow diagram for this routine. A description of this routine is provided below:

```
1  void CFixed_31_Band_Peaking::execute_filter_block_in_place(double *in)
2  {
3      int i;
4      double input = *in;
5      stages[0]->x = input;
6      stages[0]->execute_filter_stage();
7      for (i= 1; i< NUM_BANDS; i++)
8      {
9          stages[i]->x = stages[i-1]->y;
10         stages[i]->execute_filter_stage();
11     }
12     *in = stages[NUM_BANDS-1]->y;
13 }
```

- Lines 1, 2, and 13 define the boundaries of this routine.

- Line 3 defines the index variable *i* as an integer that is used to reference variables in a particular column in the matrix shown in Table 1-19.

- Line 4 defines *input* to be a value equal to the value of the sample pointed to by **in*. **in* points to the next sample to be processed by the filter section.

- Line 5 copies the sample (*input*) to the matrix variable *x* location in column 0 (the first peaking filter).

- Line 6 calls the routine *execute_filter_stage( )* which processes the input sample using the first peaking filter as noted by index *i*=0.

31-Band Graphic EQ-I

- Lines 7 through 11 contains a loop which processes the last 30 filters in cascaded fashion.

    Line 9 moves the output of the previous filter to the input of the next filter.

    Line 10 calls the routine execute_filter_stage() and processes the variables from the column indicated by the index variable i.

- Line 12 copies the output of the last filter to the memory location pointed to by *in*. The output value of the last filter is the output of the graphic equalizer. The output sample will be passed to the D/A converter or as a digital signal.

The *execute_filter_stage()* implements the second-order IIR peaking filter as shown in the network diagram in *Figure 19-10*. The coefficients and state variables are stored in the respective columns of the matrix for each filter. The variables utilized in this routine are listed below:

```
x  - Input sample
x1, x2 - past input samples
y  - Output sample
y1, y2 - past output samples
alpha, gamma, beta- filter coefficients
mu - gain coefficient
```

This routine implements the peaking filter's difference equation, Equation (19-1a), and scaled output in Equation (19-1b). The following describes this routine:

```
1 void CPeakingFilterStage::execute_filter_stage()
2 {
3     y = 2 * ((alpha * (x - x2)) + (gamma * y1) - (beta * y2));
4     x2 = x1;
5     x1 = x;
6     y2 = y1;
7     y1 = y;
8     y = (y * (mu - 1.0)) + x;
9 }
```

DSP Filters

- Lines 1, 2, and 9 define the boundaries of this routine.
- Line 3 calculates the preliminary output value y for the corresponding peaking filter.
- Lines 4 through 7 update the filter state variables (x1, x2, y1, and y2) for the corresponding peaking filter.
- Line 8 calculates the filter's scaled output value.

During shutdown of the DirectX graphic equalizer a destructor routine is used to release the memory making it available to the system for other applications. The routine ~CFixed_31_Band_Peaking() as shown below can be used for this purpose.

```
1  CFixed_31_Band_Peaking::~CFixed_31_Band_Peaking()
2  {
3       int i;
4       if (stages)
5       {
6            for (i = 0; i< NUM_BANDS; i++)
7            {
8                 delete stages[i];
9            }
10           delete stages;
11           stages = 0;
12      }
13 }
```

20

31-Band Graphic EQ-II

Introduction

This chapter presents the second of two graphic equalizer implementations. Chapter 19 utilizes the *peaking* filter presented in Chapter 10 to implement a 31-band graphic equalizer. This chapter provides step-by-step instructions to implement a 31-band graphic equalizer using the *bandpass* filter presented in Chapter 8. See the Introduction section in chapter 19 for additional background information on graphic equalizers.

Design Requirements

The goal of this design is to develop a digital 31-band graphic equalizer meeting the following specifications:

DSP Filters

Specifications:
 Channels: 2 (stereo)
 Bands: (31) with center frequencies at 1/3 octave intervals
 Gain: +/- 20 dB
 Q factor: Constant - $3\sqrt{2}$
 Sample Rate: Variable

This design implements a 31-band graphic equalizer processing two audio channels with independent processing for each channel. The graphic equalizer will support 31 bands at one-third octave intervals. Each band's filter will use a constant Q factor for predictability and minimal interaction between bands. The boost/cut range is ±20 dB with controllable step size. A graphic equalizer will be designed using bandpass filters.

Bandpass Filter Overview

Figure 20-1 displays the digital bandpass filter network. Three coefficients define the filter's characteristics with one being implemented on the feed-forward section (α) and two being implemented on the feedback section (β, and γ).

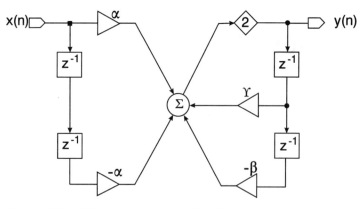

Figure 20-1. IIR bandpass shelving filter

The bandpass filter frequency response as a function of gain and phase is displayed in *Figure 20-2* at various center frequencies. The difference equation defining the bandpass filter is shown in Equation (20-1a). The coefficients for the difference equation are determined as

31-Band Graphic EQ-II

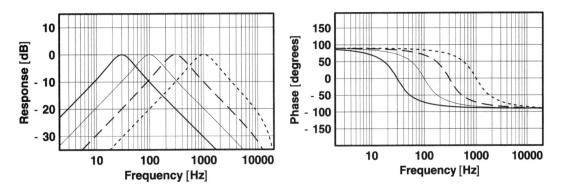

Figure 20-2. Gain and phase of digital bandpass filter with $Q = 1$ and $f_s = 44100$ Hz: solid line, $f_0 = 30$; thin solid line, $f_0 = 100$; dashed line, $f_0 = 300$; and dotted line, $f_0 = 1000$

shown in Equation (20-1b). Note that unlike the peaking filter design the coefficients in the bandpass filter are not affected by the gain. The gain is applied as a scale factor directly to the output y(n). Coefficients are affected by the quality factor (Q), center frequency (f_0), and sample rate (f_s).

The center frequency (f_0), sample rate (f_s), and quality factor (Q) each influences the calculation of the coefficients (α, β, and γ) as shown in Equation (20-1b). α, β, and γ coefficients are utilized to resolve the difference equation, Equation (20-1a), and calculate the output sample y(n). The output y(n) is then multiplied by the gain (g) to complete the

$$y(n) = 2\{\alpha\,[x(n) - x(n-2)] + \gamma\,y(n-1) - \beta\,y(n-2)\} \qquad (20\text{-}1a)$$

$$\beta = \frac{1}{2}\frac{1 - \tan[\theta_0/(2Q)]}{1 + \tan[\theta_0/(2Q)]} \qquad \gamma = \left(\tfrac{1}{2} + \beta\right)\cos\theta_0 \qquad \alpha = \left(\tfrac{1}{2} - \beta\right)/2 \qquad (20\text{-}1b)$$

$$\text{where,}\quad \theta \equiv 2\pi f / f_s \qquad \theta_0 \equiv 2\pi f_0 / f_s$$

Difference equation of the digital second-order bandpass filter with coefficient formulas

DSP Filters

implementation. Gain can also affect Q. The Q value is different when gain is positive ($Q=3\sqrt{2}$) and when it is negative ($Q = 3\sqrt{2}/4.5$). This is necessary to maintain symmetry about the zero axis.

Functional Block Diagram

The graphic equalizer consists of three basic components as shown in the system level block diagram of *Figure 20-3*. The slider section is controlled by the user and is utilized to adjust the gain (g_i) applied to each frequency band's coefficient calculation block. The function of the coefficient calculation section is to calculate the 31 sets of coefficients (α, β, and γ) applied to each of the 31 filters. The filter section utilizes the various α, β, and γ coefficients calculated for each frequency band to filter the input audio signal. The following is an overview of each section of the graphic equalizer.

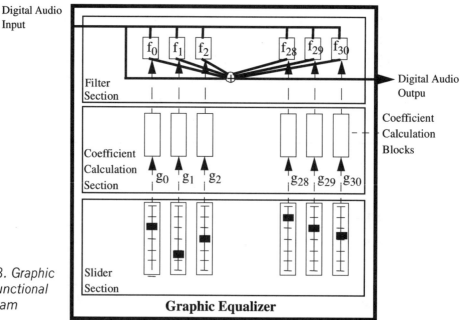

Figure 20-3. Graphic equalizer functional block diagram

Slider section

The control section of the graphic equalizer can be implemented in either hardware or in software. Regardless of the type of design required (hardware or software), the control section must provide a numerical representation of the gain controls to the coefficient calculation section.

Coefficient calculation section

The coefficient calculation section is implemented by applying the Q factor *(Q)*, center frequency *(f_o)*, and sample rate *(f_s)* values to equations 20-1b. In the implementation described below, we have chosen to apply a constant Q factor ($3\sqrt{2}$) for positive gain and $3\sqrt{2}/4.5$ for negative gain, and fixed center frequencies (one-third octaves). The remaining variables in this design are the filter gain and sample rate.

Filter section

Unlike peaking filters, bandpass filters exhibit the side effect of attenuating the stop band to a theoretical negative infinity. The bandpass filter's side bands affect adjacent frequency bands. This characteristic requires us to implement second order IIR bandpass filters in parallel. If we had chosen peaking filters we could have chosen to implement them as cascaded filters. The filter output is represented in the digital domain and must be converted to an analog signal in order for us to hear the results. The final component in a digital audio system is an audio quality D/A converter.

Flow Diagram Descriptions

The implementation of the graphic equalizer is broken into three sections displayed in the Functional Block Diagram of *Figure 20-3*. The three sections implemented are the slider section, coefficient calculation section, and filter section. The graphic equalizer implementation

DSP Filters

is first described using flow diagrams followed by detailed descriptions of the various software modules. This implementation is designed to meet the requirements as outlined in the design requirement section of this chapter.

As shown in *Figure 20-4*, there are four potential system level variables that affect the coefficients required to implement the graphic equalizer. The *sample rate* (f_s), *center frequency* (f_o), *Q factor* (Q), and *gain factor* (g) each affect the resulting coefficients used to implement the graphic equalizer. Note that both the Q factor and center frequency are fixed in this implementation.

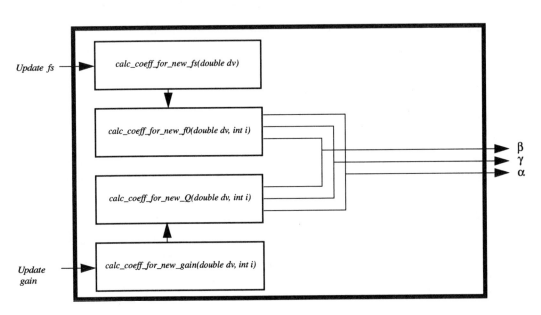

Figure 20-4. Coefficient calculation block diagram

There are two asynchronous operations taking place to implement the graphic equalizer. The first operation involves processing the audio samples in the filter section.

31-Band Graphic EQ-II

The second operation is reading the gains and updating the coefficients. The first operation is performed continuously at a fixed rate, while the second operation only occurs when a change is made to one of the variables. Therefore, these operations are asynchronous.

Updating the sample rate

When the sample rate changes, the *calc_coeff_for_new_fs(double dv)* routine is called. The flow diagram for this routine is shown in *Figure 20-5*. This routine receives only one variable *dv* that contains the new sample rate (f_s) in Hertz. The sample rate change affects all filters and therefore no index needs to be passed for this operation. The change will be applied to each filter from this routine. The *calc_coeff_for_new_f_s(double dv)* routine implements a loop where the index *i* is incremented from 0 to 30. Each iteration of the loop establishes an updated maximum center frequency based on the new sample rate and the Nyquist theorem for each filter band. *calc_coeff_for_new_f_0(double dv, int i)* is called for each iteration of the loop and updates the corresponding normalized center frequency (θ) and ultimately the respective coefficients. This routine is shown in *Figure 20-6* and described on the following page.

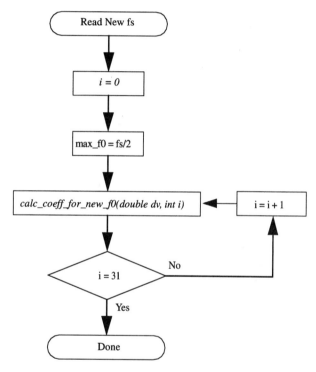

Figure 20-5. Flow diagram for updating f_s
Routine: calc_coeff_for_new_fs(double dv)

251

DSP Filters

Updating the center frequency

The calc_coeff_for_new_f_0 (double dv, int i) routine shown in Figure 20-6 is called during initialization to establish the center frequencies for each filter. It is also activated when a sample rate change occurs. Two variables are passed to this routine although the center frequency variable dv is actually a dummy variable required to conform to the format specifications in the DirectX implementation. It is ignored and has no function in this routine since the center frequencies are fixed at one-third octaves. The index i variable indicates which of the 31 filters is to be processed. The fixed center frequency (f_0) that is calculated in this routine is tested to verify that it does not exceed the Nyquist frequency (max_f_0). The normalized center frequency (θ) is calculated as a function of the center frequency and the sample rate and then the routine calc_coeff_for_new _gain(double dv, int i) is called to calculate the coefficients for the associated filter based on the new θ value.

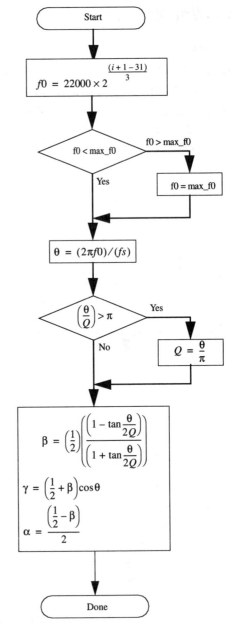

Figure 20-6. Flow diagram for updating f_0 Routine: calc_coeff_for_new_f_0(double dv, int i)

252

31-Band Graphic EQ-II

Updating the gain

The flow diagram in *Figure 20-7* represents the routine *calc_coef_for_new_gain(double dv, int i)*. This routine implements the coefficient calculation blocks. There are two variables passed to this routine; the gain (*dv*), and the index *i*. The index *i* indicates to which band the *calc_coef_for_new_gain(double dv, int i)* routine is to be applied. The variable *dv* represents the new gain value for that particular

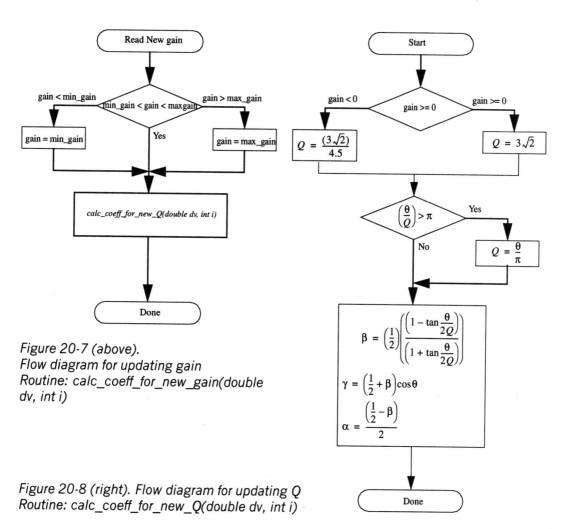

Figure 20-7 (above).
Flow diagram for updating gain
Routine: calc_coeff_for_new_gain(double dv, int i)

Figure 20-8 (right). Flow diagram for updating Q
Routine: calc_coeff_for_new_Q(double dv, int i)

DSP Filters

band. The new gain value is first tested to make sure that it falls between the maximum and minimum values established by the system for the band. If the new gain value exceeds the maximum gain value it is set equal to the maximum value. If it falls below the minimum gain value it is set equal to the minimum value. The new gain is then saved and the process to calculate the coefficients begins based on the Equation (20-1b). The routine *calc_coeff_for_new_Q(double dv, int i)* shown in the flow diagram of *Figure 20-8* is called which calculates the new coefficients using the updated gain.

Filter section

The flow diagram of *Figure 20-9* represents a routine called *execute_filter_block_in_place(double in)*. This routine implements the filter section. One variable *(in)* is passed to the routine pointing to the input sample. *execute_filter_block_in_place(double *in)* receives an input sample, processes the sample through each of the 31 cascaded peaking filters and outputs the resulting sample. The subroutine *execute_filter_stage( )* is called 31 times per sample. Each time it is called it implements one of the 31 cascaded peaking filters.

The flow diagram in *Figure 20-10* represents a routine called *execute_filter_stage( )*. *execute_filter_stage( )* implements the bandpass filter. There is only one bandpass filter routine. It is called 31 times for every new sample by the *execute_filter_block_in_place(double *in)* routine. The filter output is determined by the bandpass filter's difference equation and calculated in the *execute_filter_stage( )* routine. All of the filter states $(x_i(n-1), x_i(n-2), y_i(n-1),$ and $y_i(n-2))$ are updated and stored as well as the filter output to be used for the following filter stage. The last filter in the section outputs the resulting sample.

31-Band Graphic EQ-II

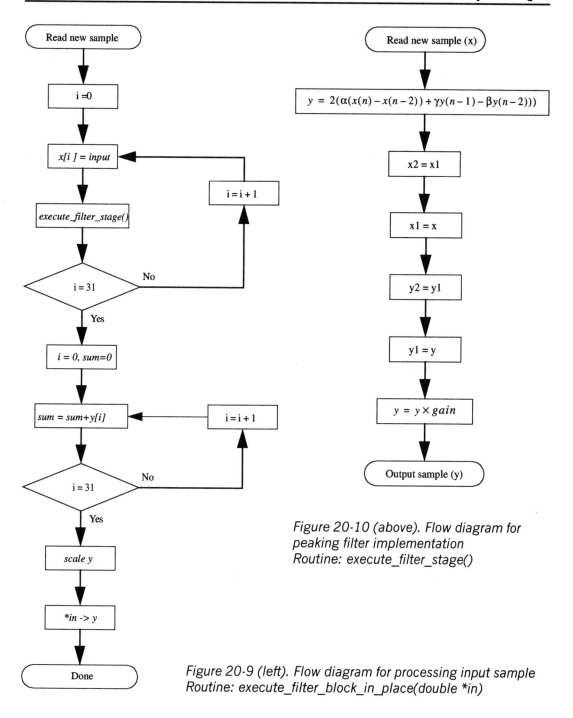

Figure 20-10 (above). Flow diagram for peaking filter implementation
Routine: execute_filter_stage()

Figure 20-9 (left). Flow diagram for processing input sample
Routine: execute_filter_block_in_place(double *in)

DSP Filters

Software Description

All filter related information is stored in a matrix as shown in Table 20-1. The matrix is used to store variables required for processing each of the bandpass filters. Each column contains the variables associated with one of the 31 bandpass filters and is indexed in the following routines by the index variable *i*. Each row represents a particular variable that is required for processing each of the bandpass filters.

Table 20-1: Peaking filter variable matrix: stages[i]->variable

Variable	i -> 0	1	2	...	...	28	29	30
min_f0 - Minimum Peaking Filter Center Frequency	0.1	0.1	0.1	....	....	0.1	0.1	0.1
f0 - Peaking Filter Center Frequency	21.5	27	34	....	....	13859	17461	22000
max_f0 - Maximum Peaking Filter Center Frequency	22049	22049	22049	....	....	22049	22049	22049
min_gain - Minimum Peaking Filter gain	-20	-20	-20	....	....	-20	-20	-20
Gain - Peaking Filter gain	0	0	0	....	....	0	0	0
max_gain - Maximum Peaking Filter gain	20	20	20	....	....	20	20	20
min_Q - Minimum Peaking Filter Quality factor	0.1	0.1	0.1	....	....	0.1	0.1	0.1
Q - Peaking Filter Quality factor	4.2	4.2	4.2	....	....	4.2	4.2	4.2
max_Q - Maximum Peaking Quality factor	20	20	20	....	....	20	20	20
x - Peaking Filter Input Sample	0	0	0	....	....	0	0	0
x1 - Previous Peaking Filter Input Sample	0	0	0	....	....	0	0	0
x2 – Previous, Previous Peaking Filter Input Sample	0	0	0	....	....	0	0	0
y - Peaking Filter Output Sample	0	0	0	....	....	0	0	0
y1 - Previous Peaking Filter Output Sample	0	0	0	....	....	0	0	0
y2 – Previous, Previous Peaking Filter Output Sample	0	0	0	....	....	0	0	0
theta0 - $2 \ast pi \ast f_0/f_s$	.00306	.00385	.00484	....	....	1.9746	2.4878	3.1345
beta - Peaking Filter feedback coefficient	.4996	.4995	.4994	....	....	.3084	.2681	.2209
gamma - Peaking Filter feedback coefficient	.9996	.9995	.9994	....	....	-.3176	-.6097	-.7209
alpha - Peaking Filter feed-forward coefficient	.0002	.0002	-.0003	....	....	.0958	.1195	.1396

The following software routines implement the functional block diagram of *Figure 20-3* as originally designed in a DirectX application and follows the program flow diagrams described in *Figures 20-4* through *20-10*. The routines are all provided as software modules written in C++.

31-Band Graphic EQ-II

Initialize the variable matrix

The implementation begins with a description of the initialization routine required to initialize the matrix to a known stable state. Initialization is performed only once prior to activation or following any system reset condition. A line-by-line analysis of the initialization routine is provided below:

```
1  CFixed_31_Band_Dual_Q::CFixed_31_Band_Dual_Q()
2  {
3      int i;
4      min_beta = -0.4999;
5      max_beta = 0.4999;
6      stages = new (CCookFilterStage *[NUM_BANDS]);
7      for (i = 0; i < NUM_BANDS; i++)
8      {
9          stages[i] = new CBandpassFilterStage;
10         stages[i]->min_f0 = 0.1;
11         stages[i]->min_gain = -0.2;
12         stages[i]->max_gain = 1.0;
13         stages[i]->min_Q = 0.1;
14         stages[i]->max_Q = 20.0;
15         stages[i]->gain = 0.0;
16         stages[i]->Q = sqrt(2.0) * 3.0;
17         stages[i]->f0 = 22000.0 * pow(2.0,((double)i+1.0-NUM_BANDS)/3.0);
18         stages[i]->max_f0 = 22049.0;
19     }
20     num_filter_stages = NUM_BANDS;
21     enabled = 1;
22 }
```

- Lines 1, 2, and 22 define the boundaries of this routine. Everything between these lines constitutes the initialization routine.

- Lines 3 through 6 define and initialize the processing parameters as follows:

    Line 3 defines the index variable i as an integer that is used to reference variables in a particular column in the matrix shown in Table 1-20. There are 31 frequency bands and thus i will operate from 0 to 30.

DSP Filters

> Lines 4 and 5 establish the minimum and maximum b coefficient values. Each time β is calculated it will be compared to these values and forced to reside between them.
>
> Line 6 calls a routine that allocates the pointers used for the variable matrix.

- Lines 7 through 19 establish a loop that will be performed NUM_BANDS times. In this example, NUM_BANDS is defined in a separate file and equals 31. This loop will initialize the matrix to a known stable states as shown in Table 1-20. Each time the loop is implemented index *i* is incremented to reference a new column in the matrix to be initialized.

  > Line 9 implements a constructor that establishes the matrix variables and associates them with the rows of the matrix.
  >
  > Lines 10 through 18 initialize each of the respective variables to the values indicated.

- Line 20 establishes num_filter_stages equal to NUM_BANDS, which in this case is 31.

- Line 21 indicates that initialization has been completed and the equalizer can now be enabled.

The following routine is a constructor that initializes the filter state variables.

```
CBandpassFilterStage::CBandpassFilterStage()
{
        x = 0;
        y = 0;
        x1 = 0;
        x2 = 0;
        y1 = 0;
        y2 = 0;
}
```

Following the completion of the initialization routine, the matrix will be initialized to the values shown in Table 1-20.

31-Band Graphic EQ-II

Coefficient calculation section

The following will describe how the coefficients are updated when a change occurs in the sample rate (f_s), center frequency (f_o) for any of the frequency bands, Q factor (Q), or gain change (g) in any channel.

Although a user interface to control sample rate is not usually provided, automatic or manual sample rate changes can be implemented in an application using the routine *calc_coef_for_new_fs (double dv)*. This routine changes the maximum center frequency (*max_f_o*) value to make sure that none of the center frequencies violates the Nyquist sampling theorem based on the new sample rate. Coefficient changes associated with sample rate are calculated in the routine *calc_coef_for_new_f_o(double dv, int i)* for each frequency band. Note that this routine modifies variables across every column of the matrix. Therefore, no index is required for this routine; however, a loop is performed that updates each column of the matrix affecting coefficients in all bands.

```
1  void CFixed_31_Band_Dual_Q::calc_coeff_for_new_fs (double dv)
2  {
3      int i;
4      fs = dv;
5      for (i = 0; i< NUM_BANDS; i++)
6      {
7          stages[i]->max_f0 = fs/2.0;
8          calc_coeff_for_new_f0 (0.0, i);
9      }
10 }
```

- Lines 1, 2, and 10 define the boundaries of this routine. The routine receives one variable *dv*. The new sample rate is denoted by the variable name *dv* and is passed in units of Hertz.
- Line 3 defines the index variable *i* as an integer that is used to reference variables in a particular column in the matrix shown in Table 20-1. There are 31 frequency bands and thus *i* will operate from 0 to 30. Each iteration is referred to as a stage in the graphic equalizer.

DSP Filters

Each stage implements one bandpass filter. Each filter has a dedicated column in the matrix allocated for its use.

- Line 4 saves the new sample rate *dv* as the variable f_s.
- Line 5 begins a loop that will process 31 iterations, one for each bandpass filter in the filter section.
- Lines 6 through 9 perform the function of modifying matrix variables.

    Line 7 calculates a new maximum center frequency (max_f_0) value based on Nyquist's theorem. The maximum center frequency should not exceed $f_s/2$.

    Line 8 calls the routine calc_coeff_for_new_f_0 (0.0,i) that updates additional variables in the matrix related to sample rate.

The routine *calc_coeff_for_new_f_0 (double dv, int i)* is called 31 times when the sample rate changes to alter the coefficients for each of the frequency bands. The index *i* indicates which of the 31 bandpass filters are being updated. *dv* is a dummy variable acting as a place saver to operate in the DirectX audio system. The value passed as *dv* is not used by the *calc_coeff_for_new_f_0 (double dv, int i)* routine. This routine can be modified to allow changing of the center frequency; however, our implementation does not provide for this feature. Adding this feature would be a good exercise. A description of this routine is provided here:

```
1  void CFixed_31_Band_Dual_Q::calc_coeff_for_new_f0 (double dv, int i)
2  {
3       CCookFilterStage *st = stages[i];
4       st->f0 = 22000.0 * pow(2.0,((double)i+1.0-NUM_BANDS)/3.0);
5       if (st->f0 > st->max_f0)
6       {
7            st->f0 = st->max_f0;
8       }
9       st->theta0 = TWO_PI * st->f0/fs;
10      if ((st->theta0 /st->Q) > PI) st->Q = st->theta0/PI ;
11      st->beta = 0.5 * (1.0 - tan(st->theta0/(2.0*st->Q))) /
                     (1.0 + tan(st->theta0/(2.0*st->Q)));
12      st->gamma = (0.5 + st->beta) * cos(st->theta0);
13      st->alpha = (0.5 - st->beta)/ 2.0;
14 }
```

- Lines 1, 2, and 14 define the boundaries of this routine.

- Line 3 establishes *st* as a pointer to the variables in column *i* of the variables matrix.

- Line 4 calculates a new center frequency for the peaking filter associated with the column being operated on.

- Lines 5 through 8 tests the center frequency verifying that it does not exceed the maximum center frequency boundaries *(max_f_0)*.

- Line 9 calculates a new normalized center frequency (θ) based on the center frequency and the updated sample rate for the associated filter.

- Line 10 updates the Q factor *(Q)* to θ/π when θ is greater than the value π.

- Line 11 calculates a new ß value based on the Equation (20-2b).

- Line 12 calculates a new γ value based on the Equation (20-2b).

- Line 13 calculates a new α value based on the Equation (20-2b).

The other two variables that affect the coefficients are gain and quality factor. These two variables are tied together as shown in the following routines. The routine *calc_coeff_for_new_gain(double dv, int i)* is called when a gain slider is changed to modify the gain of any of the 31 bands. The gain value is passed as the variable *dv* while the channel associated with the gain changed is indicated by the index variable *i*. The *calc_coeff_for_new_gain(double dv, int i)* routine does not directly change the coefficient. Instead it changes relevant variables in the respective matrix column and calls the *calc_coeff_for_new_Q(double dv, int i)* routine. A description of these routines is provided on the following page:

DSP Filters

```
1  void CFixed_31_Band_Dual_Q::calc_coeff_for_new_gain (double dv, int i)
2  {
3      CCookFilterStage *st = stages[i];
4      if (dv > st->max_gain) dv = st->max_gain;
5      else if (dv < st->min_gain) dv = st->min_gain;
6      st->gain = dv;
7      calc_coeff_for_new_Q (0.0,i);
8  }
```

- Lines 1, 2, and 8 define the boundaries for this routine.
- Line 3 establishes *st* as a pointer to the variables in column *i*.
- Line 4 tests the new gain verifying that it does not exceed the maximum gain boundaries *(max_gain)*.
- Line 5 tests the new gain verifying that it is not less than the minimum gain boundaries *(min_gain)*.
- Line 6 saves the new gain *dv* as the gain variable in column *i* of the matrix.
- Line 7 calls the *calc_coeff_for_new_Q(double dv, int i)* routine which calculates the new coefficients α, β, and γ based on the new gain value. *dv* is a dummy variable acting as a place saver in the *calc_coeff_for_new_Q(double dv, int i)* routine.

The value passed as *dv* is not used by the *calc_coeff_for_new_Q (double dv ,int i)* routine. The variable *dv* is a place saver for implementations that may provide for Q factor changes. This routine receives an index (*i*) representing the column associated with the particular filter being operated on.

```
1  void CFixed_31_Band_Dual_Q::calc_coeff_for_new_Q (double dv, int i)
2  {
3      CCookFilterStage *st = stages[i];
4      dv = (st->gain >= 0)? sqrt(2.0)*3.0 : (sqrt(2.0)/4.5)*3.0;
5      if ((st->theta0 /dv) > PI) dv = st->theta0/PI ;
```

```
6       st->Q = dv;
7       st->beta = 0.5 * (1.0 - tan(st->theta0/(2.0*st->Q))) /
                (1.0 + tan(st->theta0/(2.0*st->Q)));
8       st->gamma = (0.5 + st->beta) * cos(st->theta0);
9       st->alpha = (0.5 - st->beta)/ 2.0;
10 }
```

- Lines 1, 2, and 10 define the boundaries of this routine.

- Line 3 establishes *st* as a pointer to the variables in column *i* of the variables matrix.

- Line 4 calculates a new Q factor *(Q)* based on whether the gain is positive or negative. If the gain is zero or positive, $Q = 3\sqrt{2}$. If the gain is negative, $Q = (3\sqrt{2}/4.5)$. This is required to generate symmetrical filters about the zero axis. The parallel EQ network (shown in *Figure 20-3*), which uses constant Q bandpass filters from Chapter 8, results in an unsymmetrical frequency response without this compensation. To overcome this problem, a serial network of peaking filters (from Chapter 10) can be used to build the EQ as implemented in Chapter 19. Alternatively, we can simply use two Q factors in the parallel network of bandpass filters, one for a positive gain (boost), and a smaller Q value for the negative gain (cut). It turns out that reasonable values for a 31-band EQ are $Q = 3\sqrt{2}$ for the boost condition, and $Q = 3\sqrt{2}/4.5$ for the cut condition.

- Line 5 calculates a new Q factor based on the ratio of the new normalized center frequency θ to the present Q factor for the respective column.

- Line 6 saves the new Q factor as the Q factor variable in column *i* of the matrix.

- Line 7 calculates a new β value based on the Equation (20-2b).

- Line 8 calculates a new γ value based on the Equation (20-2b).

- Line 9 calculates a new α value based on the Equation (20-2b).

DSP Filters

Implement the peaking filter

The filter section applies the coefficients to the filter's difference equation, Equation (20-1a). This is performed in the routine *execute_filter_block_ in_place(double *in)*. *Figure 20-9* displays the flow diagram for this routine. A description of this routine is provided below:

```
1  void CFixed_31_Band_Dual_Q::execute_filter_block_in_place(double *in)
2  {
3      int i;
4      double sum = 0.0;
5      double input = *in;
6      for (i= 0; i< NUM_BANDS; i++)
7      {
8          stages[i]->x = input;
9          stages[i]->execute_filter_stage();
10     }
11     for (i = 0; i< NUM_BANDS; i++)
12     {
13         sum += stages[i]->y;
14     }
15     sum+= 0.25 * input; // constant gain stage
16     sum *= 4.0;
17     *in = sum;
18 }
```

- Lines 1, 2, and 18 define the boundaries of this routine.
- Line 3 defines the variable *i* as an integer that is used to reference variables in a particular column in the matrix shown in Table 20-1.
- Line 4 initializes the variable sum to zero.
- Line 5 defines *input* to be a value equal to the value of the sample pointed to by **in*. **in* points to the next sample to be processed by the filter section.
- Lines 6 through 10 contain a loop that applies all of the 31 bandpass filters.

31-Band Graphic EQ-II

    Line 8 copies the input sample to the input variable x in the respective column designated by the index variable i.

    Line 9 calls the routine execute_filter_stage() to apply each filter based on the updated variables in the respective column of the variables matrix designated by the index variable i.

- Lines 11 through 14 contain a loop which sums the output of all 31 filters.

    Line 13 adds the output (y) from each of the 31 filters to the variable sum to calculate the accumulative value of all the filters.

- Line 15 scales the original input by .25 and adds it to the sum value.
- Line 16 scales the summed value by a constant (4.0).
- Line 17 replaces the input sample, pointed to by *in, with the value of the summed output. *in is thus pointing to the graphic equalizer's output sample. It will be passed as the output of the graphic equalizer.

    The *execute_filter_stage()* implements the second-order IIR *bandpass* filter as shown in the network diagram in *Figure 20-10*. The coefficients and state variables are stored in the respective columns of the matrix for each filter. The variables utilized in this routine are listed on the following page:

```
x  - Input sample
x1, x2 - past input samples
y  - Output sample
y1, y2 - past output samples
alpha prime, gamma prime, beta prime - filter coefficients
mu - gain coefficient
```

    This routine implements the difference equation, Equation (20-1a). The routine is described on the following page:

DSP Filters

```
1  void CBandpassFilterStage::execute_filter_stage()
2  {
3      y = 2 * ((alpha * (x - x2)) + (gamma * y1) - (beta * y2));
4      x2 = x1;
5      x1 = x;
6      y2 = y1;
7      y1 = y;
8      y = y * gain;
9  }
```

- Lines 1, 2, and 9 define the boundaries of this routine.
- Line 3 implements the bandpass filter difference equation, Equation (20-1a), for the corresponding bandpass filter.
- Lines 4 through 7 update the filter state variables (*x1, x2, y1,* and *y2*) for the corresponding filter.
- Line 8 calculates the filter's scaled output value incorporating the gain corresponding to the slider position for the respective band.

21

4-Band Parametric EQ

Introduction

The demands of many audio applications cannot be met by the limited controls provided by graphic equalizers. For example, it is often desirable to create a very narrow EQ curve to accentuate a narrow band of frequencies. This requires the use of a high Q factor in the filter implementation that may not be able to be achieved in a graphic equalizer. Just as common is the need to remove noise or signal at a particular frequency that may not fall exactly at one of the center frequencies of a graphical equalizer. This requires having the ability to adjust a filter's center frequency. Graphic equalizers are inadequate in these applications. However, parametric equalizers are designed to meet these types of requirements.

DSP Filters

The parametric equalizer is a processor that modifies the frequency response of an audio signal provided an independent gain control for each filter implementation just like the graphic equalizer. However, the parametric equalizer is even more versatile in that it provides for both sweepable center frequencies and adjustable Q factors. This allows the parametric equalizer to implement multiple filters that can generate a uniquely shaped frequency response curve by varying the Q factors and center frequencies. This feature provides for greater versatility and artistic freedom than is available with graphic equalizers.

The processing performed by a parametric equalizer is very similar to a graphic equalizer with a few additions. Parametric equalizers are often implemented using peaking filters and in some applications *shelving filters* are provided in the outer most bands. *Figure 21-1* displays a four-band parametric equalizer with all four bands — Low, Lower-Mid, Upper-Mid, and High — being implemented with peaking filters. The HIGH band can be switched to implement a *high-pass* shelving filter while the LOW band can be switched to implement a *low-pass* shelving filter.

The additional controls and added versatility provided by parametric equalizers are crucial to professional recording artists, engineers, and producers as they provide greater influence over the music space in which they create. As a result, high quality parametric equalizers are provided on most professional mixing consoles. A valuable component to any professional studio, the equalization capabilities of a mixing console are critical to the creative process of molding and sculpting the sounds used to construct music. For example, parametric equalizers are the tools professionals use to add effect to a recording or live performance such as boosting the bass frequencies in particular instruments or even to accentuate a vocalist that may otherwise be masked by other instruments. A professional mixing console supports multiple channels of input data such as guitars, drums, vocalists, or other instruments. Each channel or sound source is allo-

4-Band Parametric EQ

cated parametric control over its frequency contour allowing the user to artistically manipulate the sound of each instrument. Parametric equalizers generally have fewer bands than graphic equalizers because their primary purpose is to tailor a few precisely defined problem frequencies. *Figure 21-1* displays a typical parametric equalizer found on a professional mixing console.

The equalizer in *Figure 21-1* will be implemented in this chapter. It applies four filters covering four bands: HIGH band, MID-HI band, MID-LO band, and LOW band. The HIGH band shown in *Figure 21-1* applies a high-pass shelving filter with a 20 kHz cutoff frequency and a 6 dB gain. The MID-HI band applies a peaking filter with 2.5 kHz center frequency and a –2 dB gain with a low Q factor. The MID-LO band applies a peaking filter with a 375 Hz center frequency and a 3 dB gain with a moderate Q factor. The LOW band applies a low-pass

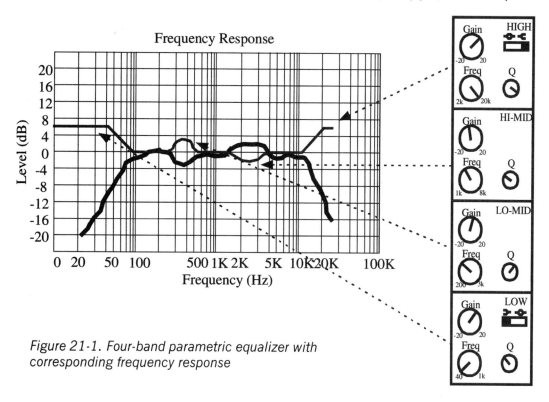

Figure 21-1. Four-band parametric equalizer with corresponding frequency response

DSP Filters

shelving filter with a 40 Hz cutoff frequency and a 6 dB gain. The center frequency control, when applied to the shelving filters, acts as the cutoff frequency control.

Much like the graphic equalizer, the uses of the parametric equalizer have not drastically changed since its inception, although the implementation has evolved from conventional passive and active analog designs to the present digital implementations of today. This chapter presents a solution for implementing a four-band digital parametric equalizer.

Design Requirements

The goal of this design is to develop a four-band parametric equalizer with switchable outer bands meeting the following specifications:

Specifications:
- Channels: 2 (stereo)
- Bands: 4
  - LOW band peaking switchable to shelving
  - LOWER-MID band peaking
  - UPPER-MID band peaking
  - HIGH band peaking switchable to shelving
- Center Frequency Ranges:
  - LOW band – 0.1Hz to 1kHz
  - LOWER-MID band – 200Hz to 5kHz
  - UPPER-MID – 2kHz to 10kHz
  - HIGH" band – 3kHz to 16kHz
- Gain: +/- 20 dB
- Q factor: 0.1 to 20
- Sample Rate: Variable

This design supports stereo (two channels) audio signals with independent processing for each channel. Each channel will be processed with four bands of parametric equalization having the outer bands switchable to shelving filters. The boost/cut range for each filter is +/–20 dB with controllable step size. The peaking filter's Q factor is adjustable for each of the bands with a range from 0.1 to 20.

Filter Overview

The following is a review of the peaking filter and shelving filter presented in Section I. Equations for implementing a peaking filter and shelving filter are provided here for convenience along with the corresponding frequency and phase response curves.

Peaking filter overview

A second-order IIR peaking filter can be implemented by summing the input $x(n)$ with the output of a second-order bandpass filter scaled by $\mu\text{-}1$, as shown in *Figure 21-2*. The bandpass output scale factor is chosen so that when $\mu = 1$, the output is equal to the input, $y(n) = x(n)$. The coefficient depends on the peaking level g, as $\mu \equiv 10^{g/20}$, where g is the boost/cut gain in dB.

The frequency and phase response of peaking filter is displayed in *Figure 21-3*. Note that since the network is second order, the phase excursion will not exceed ±90°. The maximum and minimum phase value is controlled by the peaking filter boost/cut gain factor g.

The difference equation describing a peaking filter is provided in Equation (21-1a), while the coefficients α, β, and γ are determined by Equation (21-1c).

Low-pass shelving filter overview

A first-order IIR low-pass shelving filter can be implemented by summing the input $x(n)$ with the output of a first-order low-pass filter scaled by $\mu\text{-}1$, as shown in Fig-

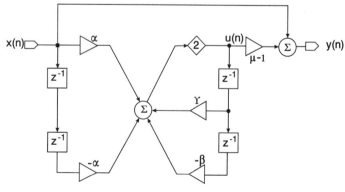

Figure 21-2. Digital peaking filter network

DSP Filters

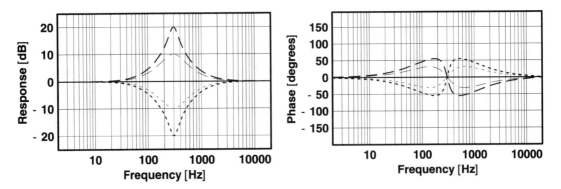

Figure 21-3. Gain and phase of digital peaking filter with Q = 1, f_0 = 300, and f_s = 44100: dotted line, g = -20 dB; thin dotted line, g = -10 dB; solid line, g = 0 dB; thin dashed line, g = +10 dB; and dashed line, g = +20 dB

ure 21-4. The low-pass output scale factor is chosen so that when $\mu = 1$, the output is equal to the input, $y(n) = x(n)$. The coefficient depends on the shelving level γ, as $\mu \equiv 10^{g/20}$, where γ is the boost/cut gain in dB.

$$u(n) = 2\{\alpha\,[x(n) - x(n-2)] + \gamma\,u(n-1) - \beta\,u(n-2)\} \qquad (21\text{-}1a)$$

$$y(n) = x(n) + (\mu - 1)\,u(n) \qquad (21\text{-}1b)$$

$$\beta = \frac{1}{2}\left(\frac{1 - \left(\frac{4}{1+\mu}\right)\tan\frac{\theta_0}{2Q}}{1 + \left(\frac{4}{1+\mu}\right)\tan\frac{\theta_0}{2Q}}\right) \qquad \gamma = \left(\tfrac{1}{2} + \beta\right)\cos\theta_0 \qquad \alpha = \left(\tfrac{1}{2} - \beta\right)/2 \qquad (21\text{-}1c)$$

$$\text{where,} \quad \theta \equiv 2\pi f / f_s \qquad \theta_0 \equiv 2\pi f_0 / f_s$$

Difference equation of the digital second-order peaking filter with coefficient formulas

4-Band Parametric EQ

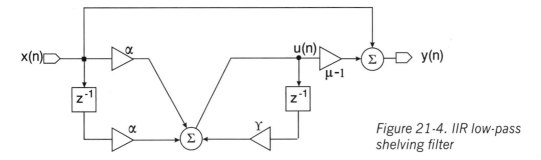

Figure 21-4. IIR low-pass shelving filter

The frequency and phase response of the low-pass shelving filter is displayed in *Figure 21-5*. Note that since the network is first order, the phase excursion will not exceed ±90°. The maximum and minimum phase value is controlled by the shelving filter boost/cut gain factor γ.

The difference equation describing a low-pass shelving filter is provided in Equation (21-2a), while the coefficients α and γ are determined by Equation (21-2c).

High-pass shelving filter overview

A first-order IIR high-pass shelving filter can be implemented by summing the input $x(n)$ with the output of a first-order high-pass filter scaled by μ-1, as shown in *Figure 21-6*. The high-pass output scale factor is cho-

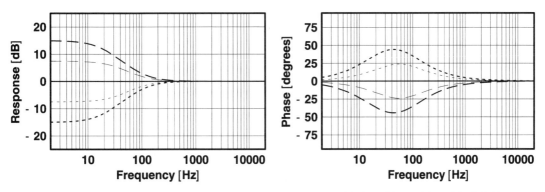

Figure 21-5. Gain and phase of low-pass shelving filter with f_c = 30 and f_s = 44100 Hz: dotted line, g = -15 dB; thin dotted line, g = -7.5 dB; solid line, g = 0 dB; thin dashed line, g = +7.5 dB; and dashed line, g = +15 dB

DSP Filters

sen so that when $\mu = 1$, the output is equal to the input, $y(n) = x(n)$. The coefficient depends on the shelving level g, as $\mu \equiv 10^{g/20}$, where g is the boost/cut gain in dB.

$$u(n) = \alpha[x(n) + x(n-1)] + \gamma u(n-1) \quad (21\text{-}2a)$$

$$y(n) = x(n) + (\mu - 1)u(n) \quad (21\text{-}2b)$$

$$\gamma = \frac{1 - \left(\dfrac{4}{1+\mu}\right)\tan\dfrac{\theta_c}{2}}{1 + \left(\dfrac{4}{1+\mu}\right)\tan\dfrac{\theta_c}{2}} \qquad \alpha = (1-\gamma)/2 \quad (21\text{-}2c)$$

where, $\theta \equiv 2\pi f / f_s \qquad \theta_c \equiv 2\pi f_c / f_s$

Difference equation of the digital first-order low-pass shelving filter with coefficient formulas

The frequency and phase response of the high-pass shelving filter is displayed in *Figure 21-7*. Note that since the network is first order, the phase excursion will not exceed $\pm 90°$. The maximum and minimum phase value is controlled by the shelving filter boost/cut gain factor g.

The difference equation describing a high-pass shelving filter is provided in Equation (21-3a), while the coefficients α and γ are determined by Equation (21-3c).

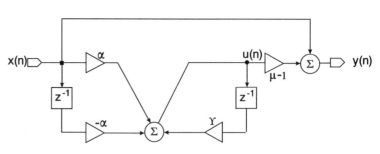

Figure 21-6. IIR high-pass shelving filter

274

4-Band Parametric EQ

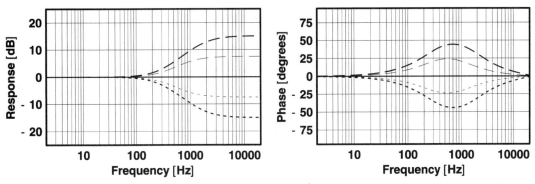

Figure 21-7. Gain and phase of high-pass shelving filter with f_c = 1000 and f_s = 44100 Hz: dotted line, g = -15 dB; thin dotted line, g = -7.5 dB; solid line, g = 0 dB; thin dashed line, g = +7.5 dB; dashed line, g = +15 dB

Equations (21-3a), (21-3b), and (21-3c) will be used to implement the high-pass shelving filters required for the parametric equalizer. Equation (21-3a) is the high-pass filter difference equation, while Equation (21-3b) incorporates the shelving component. The equations in Equation (21-3c) determine the filter coefficients (α and γ) based on the sample rate, cutoff frequency, and gain factor. Note that $\mu \equiv 10^{g/20}$.

$$u(n) = \alpha\left[x(n) - x(n-1)\right] + \gamma\, u(n-1) \quad (21\text{-}3a)$$

$$y(n) = x(n) + (\mu - 1)\, u(n) \quad (21\text{-}3b)$$

$$\gamma = \frac{1 - \left(\dfrac{1+\mu}{4}\right)\tan\dfrac{\theta_c}{2}}{1 + \left(\dfrac{1+\mu}{4}\right)\tan\dfrac{\theta_c}{2}} \qquad \alpha = (1+\gamma)/2 \quad (21\text{-}3c)$$

$$\text{where,}\quad \theta \equiv 2\pi f / f_s \qquad \theta_c \equiv 2\pi f_c / f_s$$

Difference equation of digital first-order high-pass shelving filter with coefficient formulas

Functional Block Diagram

The parametric equalizer consists of three basic components as shown in the system level block diagram of *Figure 21-8*. The control section of a parametric equalizer consists of control knobs utilized to adjust the gain factor, center frequency/cutoff frequency, and Q factor applied to each frequency band's coefficient calculation block. A switch in the HIGH and LOW bands determines whether they operate as a peaking filter or a shelving filter. When the switch is set to peaking filter the controls manipulate center frequency and Q factor. When the switch is set to shelving filter the frequency control manipulates cutoff frequency while the Q factor control has no effect. The frequency control will generically be referred to as center frequency control; however, note that when applied to the shelving filters this control actually affects the cutoff frequency. The function of the four coefficient calculation blocks is to calculate the coefficients (α, β, and γ) applied to each of the filters based on the control information read from the control section. The filter section utilizes the various α, β, and γ coefficients calculated for each frequency band to filter the input audio signal. The coefficients are applied to the peaking and shelving filters realized as direct form I implementations. When the HIGH or LOW band switches are set for shelving filter operation, these bands' peaking filters can be set to pass the audio unaffected and vice versa. The following outlines the three sections shown in *Figure 21-8*.

Control section

The control section of the parametric equalizer can be implemented in either hardware or in software. Regardless of the type of design required (hardware or software), the control section must provide a numerical representation of all controls to the coefficient calculation section.

4-Band Parametric EQ

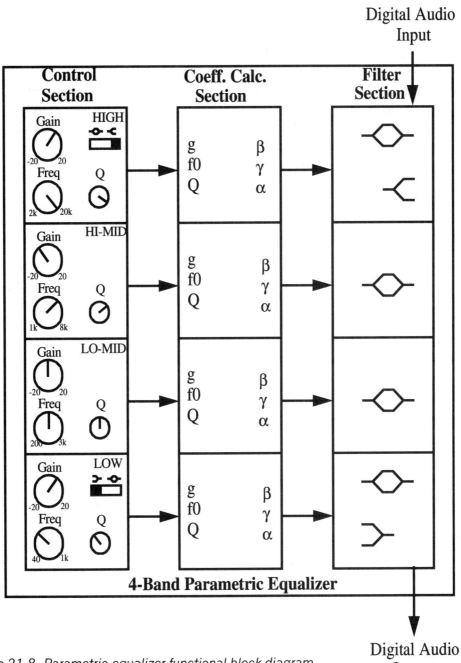

Figure 21-8. Parametric equalizer functional block diagram

Coefficient calculation section

The coefficient calculation section is implemented by applying the gain, center frequency, and Q factor values read from the control panel to the equations derived in the peaking filter and shelving filter chapters. There are several variables, as shown by the respective set of filter equations, that affect the peaking and shelving filter operation including sample rate, center frequency, Q factor, and gain.

Filter section

As mentioned in previous chapters, audio signals are by nature analog and must be converted to a numerical representation by an A/D converter to be filtered in the digital domain. Following A/D conversion, filtering is applied in series in the digital domain. Four filters are applied in series with the first LOW filter being applied as either a second-order peaking filter or a first-order low-pass shelving filter. Two second-order peaking filters are applied for the MID-HI and MID-LO filters with the last HI filter being either a second-order peaking filter or a first-order high-pass shelving filter. The output of the last filter provides the output of the parametric equalizer. This output must be converted back to the analog domain for the results to be heard by the user.

Flow Diagram Descriptions

The following flow diagrams describe the implementation of software to realize the four-band parametric equalizer. As shown in *Figure 21-9*, the coefficient calculation section utilizes four routines to determine the coefficients α, β, and γ. When the Q factor is adjusted in the control section for any of the four bands, the routine *calc_coef_for_new_Q(double dv, int i)* is activated to update the new Q factor value in the matrix table. This routine in turn calls the *calc_coef_for_new_gain(double dv, int i)* routine to complete the coefficient calculation process utilizing the update Q factor. Similarly, when the center frequency control is adjusted in the control section for any of the four bands, the routine *calc_coef_for_new_f0(double dv, int i)* is activated to update the new center frequency value in the matrix table. The *calc_coef_for_new_gain(double dv, int i)* routine is called to complete the calculation process using the updated center frequency value. Adjustments in one of the gain controls will activate the *calc_coef_for_new_gain(double dv, int i)* routine, which will directly calculate new coefficients for the corresponding filter.

Note the index (*int i*) that is passed as a variable in each of these routines. This index indicates to which band the adjustment should be applied. Sample rate also affects the coefficient calculation. A sample rate change will activate the *calc_coef_for_new_fs(double dv)* routine. Note that since the sample rate affects the coefficient calculations in each band there is no index passed. This change is applied to each

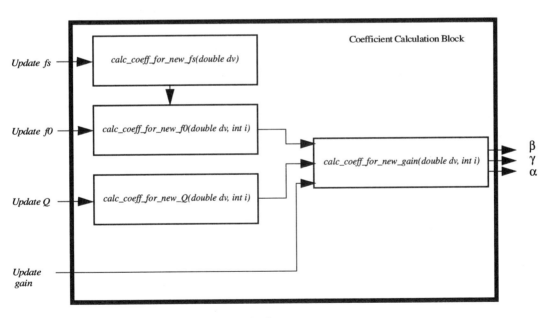

Figure 21-9. Coefficient calculation block diagram

DSP Filters

filter implementation. The sample rate affects the value of the normalized center frequency (θ) for each band, which is calculated in the *calc_coef_for_new_f0(double dv, int i)* routine. Thus, the *calc_coef_for_new_fs(double dv)* routine updates the sample rate values in the matrix and then activates the *calc_coef_for_new_f0(double dv, int i)* routine to calculate the new θ for each band. The routine *calc_coef_for_new_gain(double dv, int i)* is then called to complete the coefficient calculation process given the updated θ value. A more detailed description is provided with the flow diagrams.

Updating the Q factor

The flow diagram in *Figure 21-10* shows the operation of the *calc_coef_for_new_Q(double dv, int i)* routine. The updated Q factor is passed to the routine as the variable *dv*. This routine is activated when any of the Q factor controls are adjusted in the control section. The updated Q factor is tested to verify that it falls between the system-defined limits (*min_Q, max_Q*). The Q factor is then compared with the value θ/π, and if it is found to be larger it will be clamped to the value θ/π. This is applied to compensate for the lack of filtering symmetry about the 0 dB axis. The routine *calc_coef_for_new_gain(double dv, int i)* is then activated to calculate the coefficients for the filter indicated by the index *i* using the updated Q factor value. The *calc_coef_for_new_gain(double dv, int i)* routine is described later in this chapter.

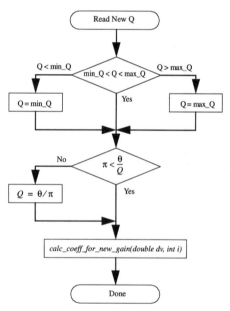

Figure 21-10.
Flow diagram for updating Q factor
Routine: calc_coeff_for_new_Q(double dv, int i)

4-Band Parametric EQ

Updating the sample rate

When the sample rate of the input data changes, the *calc_coeff_for_new_fs(double dv)* routine is called as shown in *Figure 21-11*. This routine receives only one variable *dv* that contains the new sample rate in Hertz. The sample rate change affects all filters and therefore no index *i* is required for this operation. The change will be applied to each filter from this routine. The *calc_coeff_for_new_fs (double dv)* routine implements a loop where the index *i* is incremented from 0 to 5. Each iteration of the loop establishes an updated maximum center frequency based on the new sample rate and the Nyquist theorem for each filter band. *calc_coeff_for_new_f_0(double dv, int i)* is called for each iteration of the loop and updates the corresponding θ values and, ultimately, the respective coefficients. This routine is shown in *Figure 21-12* and described below.

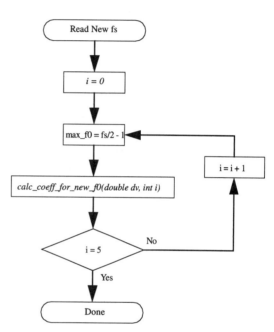

Figure 21-11. Flow diagram for updating fs Routine: calc_coeff_for_new_fs(double dv)

Updating the center frequency

A change in either sample rate or center frequency activates the routine *calc_coeff_for_new_f_0(double dv, int i)*. Two variables are passed to this routine—the center frequency *dv* and the index *i*. The updated center frequency is tested to verify that it falls between the system-defined limits (*min_f_0, max_f_0*). θ is calculated as a function of the center

DSP Filters

frequency and the sample rate and then the routine *calc_coeff_for_new_gain(double dv, int i)* is called to calculate the coefficients based on the new θ value.

Updating the gain

The *calc_coeff_for_new_gain(double dv, int i)* routine is where the coefficients are calculated based on changes in sample rate, center frequency, Q factor or gain in any of the filter bands. This routine implements the flow diagram in *Figure 21-13* and is called any time one of the variables is changed in any of the filter bands. Two variables are passed to this routine—the gain (*dv*) and the index *i*. In the event that the gain has been updated in any filter band the gain is tested to verify that it falls between the system-defined limits (*min_gain, max_gain*). The gain scale factor (*mu*) is calculated as a function of the gain. The coefficients are to be determined based on the type of filter to be implemented. The low-pass shelving filter is applied when the index value is 0. Peaking filters are applied when the index value is 1, 2, 3, or 4. The high-pass shelving filter is applied when the index value is 5.

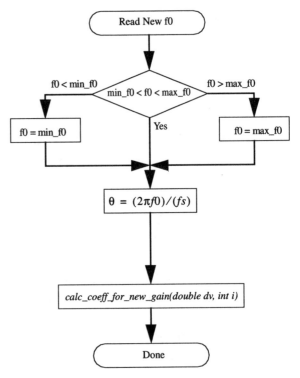

Figure 21-12. Flow diagram for updating f_0
Routine: calc_coeff_for_new_f_0(double dv, int i)

4-Band Parametric EQ

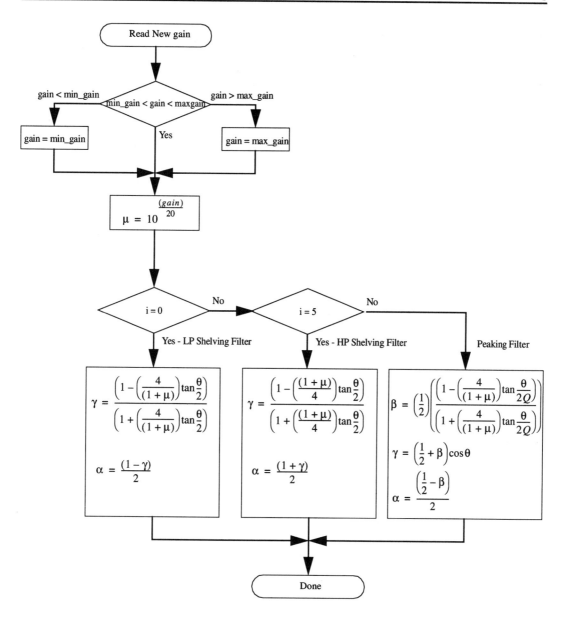

Figure 21-13. Flow diagram for updating gain
Routine: calc_coeff_for_new_gain(double dv, int i)

DSP Filters

Software Description

All filter related information is stored in a two-dimensional array or matrix as shown in Table 1-21. The matrix is used to store variables required for processing each of the peaking filters. Each column contains the variables associated with one of the six filters in the filter section and is referenced in the following routines by the value i. Each row represents a particular variable that is required for processing each of the peaking filters.

The following software routines implement the functional block diagram of *Figure 21-8* as originally designed in a DirectX application and follow the program flow diagrams described in *Figures 21-9* through *21-13*. The routines are all provided as software modules written in C++.

The implementation begins with the initialization routine required to initialize the variable array to a known stable state prior to activation of the equalizer. Initialization is performed only once prior to activation or following any system reset condition.

Table 21-1: Peaking filter variable array: stages[i]->variable

Variable	i ->	0 Shelving	1 Peaking	2 Peaking	3 Peaking	4 Peaking	5 Shelving
min_f0 - Minimum Peaking Filter Center Frequency		0.1	0.1	0.1	0.1	0.1	0.1
f0 - Peaking Filter Center Frequency		300	200	621.49	1930.97	6000	4000
max_f0 - Maximum Peaking Filter Center Frequency		22049	22049	22049	22049	22049	22049
min_gain - Minimum Peaking Filter gain		-20	-20	-20	-20	-20	-20
Gain - Peaking Filter gain		0	0	0	0	0	0
max_gain - Maximum Peaking Filter gain		20	20	20	20	20	20
min_Q - Minimum Peaking Filter Q factor		0.1	0.1	0.1	0.1	0.1	0.1
Q - Peaking Filter Q factor		1.0	1.4	1.4	1.4	1.4	1.0
max_Q - Maximum Peaking Q factor		20	20	20	20	20	20
x - Peaking Filter Input Sample		0	0	0	0	0	0
x1 - Previous Peaking Filter Input Sample		0	0	0	0	0	0
x2 – Previous, Previous Peaking Filter Input Sample		-	0	0	0	0	-
y - Peaking Filter Output Sample		0	0	0	0	0	0
y1 - Previous Peaking Filter Output Sample		0	0	0	0	0	0
y2 – Previous, Previous Peaking Filter Output Sample		-	0	0	0	0	-
mu - $10^{G/20}$		1	1	1	1	1	1
theta0 - $2*pi*f_0/f_s$		.00306	.00385	.00484	1.9746	2.4878	3.1345
beta - Peaking Filter feedback coefficient		.4996	.4995	.4994	.3084	.2681	.2209
gamma - Peaking Filter feedback coefficient		-.6277	-.6277	-.6277	-.6277	-.6277	-.6277
alpha - Peaking Filter feed-forward coefficient		-.6277	-.6277	-.6277	-.6277	-.6277	-.6277

4-Band Parametric EQ

Initializing the variable matrix

The initialization routine establishes several parameters necessary to allow the equalizer to meet the goals defined at the outset of this section. A line-by-line analysis of the initialization routine is provided.

```
1  CParametric_1::CParametric_1()
2  {
3      int i;
4      double f1 = 200.0;
5      double f2 = 6000.0;
6      double Df;

7      min_beta = -0.4999;
8      max_beta = 0.4999;
9      stages = new (CCookFilterStage * [NUM_BANDS]);
10     stages[0] = new CLP_ShelvingFilterStage;
11     stages[NUM_BANDS-1] = new CHP_ShelvingFilterStage;

12     for (i = 1; i < NUM_BANDS-1; i++)
13     {
14         stages[i] = new CPeakingFilterStage;
15     }

       // Initialization of LP filter
16     {
17     stages[0]->min_f0 = 0.1;
18     stages[0]->min_gain = -20.0; //db
19     stages[0]->max_gain = +20.0;
20     stages[0]->min_Q = 0.1;
21     stages[0]->max_Q = 20.0;
22     stages[0]->gain = 0.0;
23     stages[0]->Q =1.0;
24     stages[0]->f0 = 300.0;
25     stages[0]->max_f0 = 22049.0;
26     }

27     Df=pow(10,(log10(f2)-log10(f1))/(NUM_BANDS-3));
28
       // Initialization of Peaking filters
29     for (i = 1; i < NUM_BANDS-1; i++)
30     {
```

DSP Filters

```
31              stages[i]->min_f0 = 0.1;
32              stages[i]->min_gain = -20.;
33              stages[i]->max_gain = 20.0;
34              stages[i]->min_Q = 0.1;
35              stages[i]->max_Q = 20.0;
36              stages[i]->gain = 0.0;
37              stages[i]->Q = sqrt(2.0);
38              stages[i]->f0 = f1 * pow(Df,((double)i-1));
39              stages[i]->max_f0 = 22049.0;
40      }

        // Initialization of HP filters
41      {
42              stages[NUM_BANDS-1]->min_f0 = 0.1;
43              stages[NUM_BANDS-1]->min_gain = -20.0;   //db
44              stages[NUM_BANDS-1]->max_gain = +20.0;
45              stages[NUM_BANDS-1]->min_Q = 0.1;
46              stages[NUM_BANDS-1]->max_Q = 20.0;
47              stages[NUM_BANDS-1]->gain = 0.0;
48              stages[NUM_BANDS-1]->Q =1.0;
49              stages[NUM_BANDS-1]->f0 = 4000.0;
50              stages[NUM_BANDS-1]->max_f0 = 22049.0;
51      }

52      num_filter_stages = NUM_BANDS;
53      enabled = 1;

54 }
```

- Lines 1, 2, and 54 define the boundaries of this routine. Everything between these lines constitutes the initialization routine.

- Lines 3 through 8 define and initialize the processing parameters as follows:

    Line 3 defines the index i that is used to reference variables in a particular column in the matrix. There are six columns, one for each filter. Thus, i will continually increment from 0 to 5. After reaching 5, i will be reset to 0.

    Lines 4 and 5 initialize the low-pass and high-pass shelving filter cutoff frequencies as 200Hz and 6kHz respectively.

    Line 6 defines the variable Df that is used to calculate the center frequencies for the four peaking filters.

4-Band Parametric EQ

Lines 7 and 8 establish the minimum and maximum β coefficient values. See Section 8 for a definition of β. Each time β is calculated it will be compared to these values and forced to reside between them.

- Line 9 calls a routine that clears the memory used for the matrix.

- Line 10 implements a constructor *CLP_ShelvingFilterStage* that establishes the low-pass shelving filter matrix variables and associates them with the corresponding rows of the matrix. The *CLP_ShelvingFilterStage* constructor is shown below.

- Line 11 implements a constructor *CHP_ShelvingFilterStage* that establishes the high-pass shelving filter matrix variables and associates them with the corresponding rows of the matrix. The *CHP_ShelvingFilterStage* constructor is shown on the following page.

- Lines 12 through 15 are a loop that implements a constructor *CPeakingFilterStage* establishing the four peaking filter matrix variables and associates them with the corresponding rows of the matrix. The *CPeakingFilterStage* constructor is shown on the following page.

- Lines 16 thorugh 26 initialize the low-pass shelving filter variables with the values indicated.

- Line 27 calculates a constant *Df* based on the LOW and HIGH band center frequencies used to initialize the MID-LOW and MID-HI band center frequencies in line 38.

- Lines 29 through 40 define a loop that initializes the peaking filter variables with the values indicated.

- Lines 41 through 51 initialize the high-pass shelving filter variables with the values indicated.

- Line 52 establishes num_filter_stages equal to NUM_BANDS, which in this case is 6.

- Line 53 indicates that initialization has been completed and the equalizer can now be enabled.

DSP Filters

CPeakingFilterStage constructor:

```
1  CPeakingFilterStage::CPeakingFilterStage()
2  {
3       x = 0;
4       y = 0;
5       x1 = 0;
6       x2 = 0;
7       y1 = 0;
8       y2 = 0;
9  }
```

CHP_ShelvingFilterStage constructor:

```
1  CHP_ShelvingFilterStage::CHP_ShelvingFilterStage()
2  {
3       x = 0;
4       y = 0;
5       x1 = 0;
6       y1 = 0;
7  }
```

CLP_ShelvingFilterStage constructor:

```
1  CLP_ShelvingFilterStage::CLP_ShelvingFilterStage()
2  {
3       x = 0;
4       y = 0;
5       x1 = 0;
6       y1 = 0;
7  }
```

The three constructors above are used during system initialization to clear the variables indicated for each column in the matrix.

Calculating the coefficients

When the gains, center frequencies, Q factors, and sample rate values have been updated for each of the filters, the coefficients can be calculated. Each filter's gain, center frequency, and Q factor are adjust-

4-Band Parametric EQ

able by the user. The routines provided in this chapter also provide for sample rate changes. The following is an implementation of the block diagram shown in *Figure 21-9*.

This routine implements the flow diagram shown in *Figure 21-10*.

```
1  void CParametric_1::calc_coeff_for_new_Q (double dv, int i)
2  {
3       // Only recalc for change in Q on peak filters.
4       if (i >= 1 && i < NUM_BANDS-1)
5       {
6            CCookFilterStage *st = stages[i];

7            if (dv > st->max_Q) dv = st->max_Q;
8            else if (dv < st->min_Q) dv = st->min_Q;
9            if ((st->theta0 /dv) > PI) dv = st->theta0/PI ;
10           st->Q = dv;
11           calc_coeff_for_new_gain (st->gain, i);
12      }
13 }
```

- Lines 1, 2, and 13 define the boundaries of this routine. The routine receives two variables — Q factor *(dv)* and frequency band index *i*.
- Line 3 comment.
- Lines 4, 5, and 12 define the boundaries of an *if* statement that controls the processing to only the peaking filters in columns 1 through 4.
- Line 6 establishes *st* as a pointer to the variables in column *i* of the variables matrix.
- Line 7 tests the Q factor verifying that it does not exceed the maximum Q factor boundaries *(max_Q)*.
- Line 8 tests the Q factor verifying that it is not less than the minimum Q factor boundaries *(min_Q)*.
- Line 9 tests the value of θ/Q. If θ/Q is greater than π, then $Q = \theta/\pi$.
- Line 10 assigns the value *(dv)* to the variable Q.

DSP Filters

- Line 11 calls the routine calc_coeff_for_new_gain(st->gain, i). The gain value is pulled from the matrix in column i and passed to the new routine. This routine uses the updated Q value to calculate the coefficients ß, γ, and α.

The following routine implements the flow diagram shown in *Figure 21-11*.

```
1  void CParametric_1::calc_coeff_for_new_fs (double dv)
2  {
3      int i;
4      fs = dv;
5      for (i = 0; i< NUM_BANDS; i++)
6      {
7          stages[i]->max_f0 = fs/2.0 -1.0;
8          calc_coeff_for_new_f0 (stages[i]->f0, i);
9      }
10 }
```

- Lines 1, 2, and 10 define the boundaries of this routine. The routine receives one variable — new sample rate. The new sample rate is denoted by the variable name *dv*.

- Line 3 defines the variable *i* as an integer that is used to reference variables in a particular column in the matrix shown in Table 1-21.

- Line 4 saves the new sample rate *dv* as the variable f_s.

- Line 5 begins a loop that will process six iterations, one for each filter band.

- Lines 6 through 9 perform the function of modifying matrix variables.

    Line 7 calculates a new maximum center frequency (max_f0) value based on Nyquist's theorom. The maximum center frequency should not exceed one-half of the sample frequency.

    Line 8 calls the routine calc_coeff_for_new_f0 (0.0,i) that updates additional variables in the matrix related to sample rate.

4-Band Parametric EQ

The following routine implements the flow diagram shown in *Figure 21-12*.

```
1 void CParametric_1::calc_coeff_for_new_f0 (double dv, int i)
2 {
3      CCookFilterStage *st = stages[i];
4      if (dv > st->max_f0) dv = st->max_f0;
5      else if (dv < st->min_f0) dv = st->min_f0;
6      st->f0 = dv;
7      st->theta0 = TWO_PI * st->f0/fs;
8      calc_coeff_for_new_gain (st->gain, i);
9 }
```

- Lines 1, 2, and 9 define the boundaries of this routine. The routine receives two variables — filter center frequency (*dv*) and frequency band *i*.

- Line 3 establishes *st* as a pointer to the variables in column *i* of the variables matrix.

- Line 4 tests the center frequency verifying that it does not exceed the maximum center frequency boundaries *(max_f_0)*.

- Line 5 tests the center frequency verifying that it is not less than the minimum center frequency boundaries *(min_f_0)*.

- Line 6 copies the center frequency value *dv* passed to the routine in the f_0 location in the column indicated by *i*.

- Line 7 calculates a new θ based on the center frequency and sample rate and then saves the value in the θ location in the column indicated by *i*.

- Line 8 calls the routine *calc_coeff_for_new_gain (st->gain, i)*. The gain is passed to this routine along with the column index *i*. This routine calculates the new coefficients α, β, and γ given the updated variables based on the updated value θ.

The following routine implements the flow diagram shown in *Figure 21-13*.

DSP Filters

```
1  void CParametric_1::calc_coeff_for_new_gain (double dv, int i)
2  {
3       CCookFilterStage *st = stages[i];
4       if (dv > st->max_gain) dv = st->max_gain;
5       else if (dv < st->min_gain) dv = st->min_gain;
6       st->gain = dv;
7       st->mu = pow(10.0,st->gain/20.0);
8       if (i == (NUM_BANDS-1)) // HP shelving filter
9       {
10              st->gamma = (1.0-(((1.0+st->mu)/4.0) * tan(st->theta0/2.0)))/
                                (1.0+(((1.0+st->mu)/4.0) * tan(st->theta0/2.0)));
11              st->alpha = (1.0 + st->gamma)/2.0;
12      }
13      else if (i == 0) // LP shelving
14      {
15              st->gamma = (1.0-((4.0/(1.0+st->mu)) * tan(st->theta0/2.0)))/
                                (1.0+((4.0/(1.0+st->mu)) * tan(st->theta0/2.0)));
16              st->alpha = (1.0 - st->gamma)/2.0;
17      }
18      else // Peaking filters
19      {
20              st->beta = 0.5 * (1.0 - (4.0/(1.0+st->mu))*tan(st->theta0/(2.0*st->Q))) /        (1.0 + (4.0/(1.0+st->mu))*tan(st->theta0/(2.0*st->Q)));
21              st->gamma = (0.5 + st->beta) * cos(st->theta0);
22              st->alpha = (0.5 - st->beta)/2.0;
23      }
24 }
```

- Lines 1, 2, and 24 define the boundaries for this routine.
- Line 3 establishes *st* as a pointer to the variables in column i.
- Line 4 tests the new *gain* verifying that it does not exceed the maximum gain boundaries *(max_gain)*.
- Line 5 tests the new *gain* verifying that it is not lower than the minimum gain boundaries *(min_gain)*.
- Line 6 saves the new gain in column i of the matrix.

4-Band Parametric EQ

- Line 7 calculates a new μ value.
- Lines 8 through 12 calculate new α and γ for the high-pass shelving filter.
- Lines 13 through 17 calculate new α and γ for the low-pass shelving filter.
- Lines 18 through 23 calculate new α, β, and γ for the peaking filters.

Implementing the filters

The main filter loop is implemented with the following *execute_filter_block_in_place(double *in)* routine. The input sample is passed to the routine via the pointer **in* and is initially stored as the *x* variable for the first filter (index *i* = 0, low-pass shelving filter). The routine *execute_filter_stage()* is called to process the input sample with the low-pass shelving filter. The output of the low-pass shelving filter is stored as the variable *y*. A loop operates on the next five filters always moving the output of the previous filter into the input of the next filter, thus performing filtering in a cascade manner.

```
1  void CParametric_1::execute_filter_block_in_place(double *in)
2  {
3      int i;
4      double input = *in;
5      stages[0]->x = input;
6      stages[0]->execute_filter_stage();
7      for (i= 1; i< NUM_BANDS; i++)
8      {
9          stages[i]->x = stages[i-1]->y;
10         stages[i]->execute_filter_stage();
11     }
12     *in = stages[NUM_BANDS-1]->y;
13 }
```

- Lines 1, 2, and 13 define the boundaries of this routine.
- Line 3 defines the variable *i* as an integer that is used to reference variables in a particular column in the matrix shown in Table 1-21. There are six frequency bands. Thus, *i* will operate from 0 to 5. Each iteration is referred to as a stage in the parametric equalizer. Each

DSP Filters

stage implements one filter and has a dedicated column in the matrix allocated for its use.

- Line 4 defines a variable input initialized to the pointer value *in. *in is a pointer to the next input sample.
- Line 5 stores the input sample value in the first filter's variable x in the first column of the matrix.
- Line 6 calls the routine *execute_filter_stage( )* which processes the first filter (low-pass shelving).
- Lines 7 through 11 contain a loop that processes the last five filters in cascaded fashion.
  > Line 9 moves the output of the previous filter to the input of the present filter.
  > Line 10 calls the routine execute_filter_stage() to operate on the present filter stage variables.
- Line 12 relocates the pointer *in to the output of the last filter stage. The output value of the last filter is the output of the graphic equalizer. *in is thus pointing to the graphic equalizer's output sample. It will be passed to the D/A converter.

The *CPeakingFilterStage::execute_filter_stage()* implements the second-order IIR peaking filter as shown in *Figure 21-2*. The peaking filter coefficients and state variables are stored in the corresponding column of the matrix for each peaking filter. The variables utilized in this routine are listed below:

```
x - Input sample
x1, x2 - past input samples
y - Output sample
y1, y2 - past output samples
alpha prime, gamma prime, beta prime - filter coefficients
mu - gain coefficient
```

This routine implements the difference equation for the peaking filter shown in Equations (21-1a) and (21-1b).

4-Band Parametric EQ

```
1  void CPeakingFilterStage::execute_filter_stage()
2  {
3      y = 2 * ((alpha * (x - x2)) + (gamma * y1) - (beta * y2));
4      x2 = x1;
5      x1 = x;
6      y2 = y1;
7      y1 = y;
8      y = (y * (mu - 1.0)) + x;
9  }
```

- Lines 1, 2, and 9 define the boundaries of this routine.

- Line 3 implements the peaking filter difference equation, Equation (21-1a), for the corresponding filter.

- Lines 4 through 7 update the filter state variables (x1, x2, y1, and y2) for the corresponding filter.

- Line 8 calculates the filter's output value.

The following routine implements the high-pass shelving filter difference equation. Equation (21-3a) and (21-3b). The *CHP_ShelvingFilterStage::execute _filter_stage()* implements the first-order IIR high-pass shelving filter as shown in *Figure 21-6*. The coefficients and state variables are stored in the matrix for each stage. The variables utilized in this routine are listed here:

```
x  - Input sample
x1 - past input sample
y  - Output sample
u1 - past lpf output sample
alpha prime, gamma prime, beta prime - filter coefficients
mu - gain coefficient
```

```
1  void CHP_ShelvingFilterStage::execute_filter_stage()
2  {
3      double u;
4      u = (alpha * (x - x1) + (gamma * u1) );
5      x1 = x;
6      u1 = u;
7      y = (u * (mu - 1.0)) + x;
8  }
```

DSP Filters

- Lines 1, 2, and 8 define the boundaries of this routine.
- Line 3 implements the high-pass shelving filter difference Equation (23-1a).
- Line 4 calculates the preliminary high-pass filter output value u.
- Lines 5 and 6 update the filter state variables (x1, y1).
- Line 8 calculates the filter's output value y.

This routine implements the low-pass shelving filter difference equation, (21-2a) and (21-2b). The *CLP_ShelvingFilterStage::execute_filter_stage()* implements the first-order IIR low-pass shelving filter as shown in Figure 21-4. The coefficients and state variables are stored in the matrix for each stage. The variables utilized in this routine are listed below:

```
x  - Input sample
x1 - past input sample
y  - Output sample
u1 - past lpf output sample
alpha prime, gamma prime, beta prime - filter coefficients
mu - gain coefficient
```

```
1 void CLP_ShelvingFilterStage::execute_filter_stage()
2 {
3     double u;
4     u = (alpha * (x + x1) + (gamma * u1));
5     x1 = x;
6     u1 = u;
7     y = (u * (mu - 1.0)) + x;
8 }
```

- Lines 1, 2, and 8 define the boundaries of this routine.
- Line 3 implements the low-pass shelving filter difference Equation (22-1a).
- Line 4 calculates the preliminary low-pass filter output value u.
- Lines 5 and 6 update the filter state variables (x1, u1).
- Line 8 calculates the filter's output value y.

22

Digital Crossover

Introduction

Audio systems, whether used in professional or consumer applications, typically aim to reproduce the entire audible range of frequencies, but this is well beyond the handling capabilities of any single loudspeaker driver. If there were a single driver capable of producing the entire audible range accurately, there would be no need for this chapter. Therefore, several drivers (speaker cones, horns), each covering different frequency ranges, are combined to reproduce the full audio spectrum. Driving a speaker cone with signals outside of the frequency range for which it is designed can potentially cause both distortion to the signal and damage to the speaker cone.

Hence, it is necessary to filter the input signal into bands that can be directed to the appropriate speaker drivers. This is the function of a crossover. In *Figure 22-1* a loudspeaker with two drivers is shown — a

DSP Filters

woofer for lower frequencies and a tweeter for higher frequencies. The incoming digital signal is split into two parts: a low-pass filter section removes high frequencies and a high-pass filter section removes the low frequencies. This is a two-way crossover. The filtered signals are amplified and sent to the appropriate driver. The aim of the crossover design is to try to make the resulting acoustical output frequency response of the speaker/crossover system as flat as possible. This is the ideal speaker/crossover design. However, in practice this is a difficult task that requires knowledge of the speaker and cabinet characteristics and designing the crossover to match and compensate for certain variances.

The crossover slope should be steep enough that transient distortion is not introduced in the frequency response. Ideally, the phase response of the crossover network should be the inverse of the phase

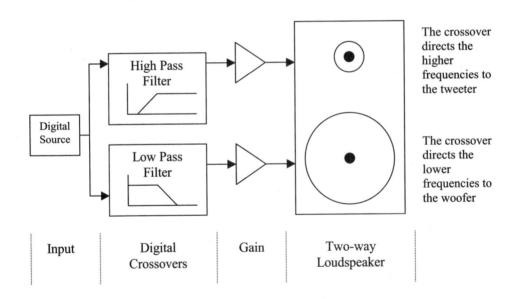

Figure 22-1. A full bandwidth digital input signal is passed through digital crossover filters where the lower frequencies are separated from the higher frequencies. The separate signals are then amplified, with the lower frequencies sent to the woofer and the higher frequencies sent to the tweeter.

Digital Crossover

response of the speaker drivers canceling any potential phase distortion. There are a number of factors that need to be considered when selecting the crossover frequency for a particular design, such as the combined on-axis and off-axis response and the frequency response capabilities of the individual speaker drivers.

Speaker crossovers are commonly implemented within speaker cabinets using simple high-voltage crossover networks with passive components (resistors, capacitors, and inductors). Most low-cost audio systems implement passive crossover networks within the speaker cabinets. This passive design is simple, inexpensive, and incorporates all required operating components in a single unit. However, this design places additional burden on the power amplifier and does not provide optimal performance. Thus, many larger or high-end systems implement separate crossovers using low-voltage active networks. Active crossovers are implemented prior to amplification of the signal. This type of system requires bi-amplification or tri-amplification (two or three separate power amplifiers, each handling a separate frequency band) for each speaker cabinet and adds cost, but also improves performance.

Many of the next-generation, high-end audio systems are incorporating digital capabilities, including digital power amplification and digital crossover networks. This maintains all processing in the digital domain up to the speaker itself. This type of design can provide both improved quality as well as reduced cost. This chapter will provide a solution for implementing a digital crossover.

Which Filters Should be Considered in Crossover Designs?

On the surface, the task at hand appears to be outlined in *Figure 22-1*, namely passing the incoming full bandwidth audio signal through two filters: one a low-pass filter and the other a high-pass filter. However,

DSP Filters

it is not clear what type of filters these should be. Let us examine the responses of a few filters to determine which ones may be appropriate. For this example, a two-way design is considered with the crossover frequency at 1 kHz. The sampling rate is assumed to be 44.1 kHz.

Second-order Butterworth filter

The most commonly used filters in the industry today are the second-order Butterworth filters. *Figure 22-2* shows the magnitude and phase responses for the low-pass and high-pass sections.

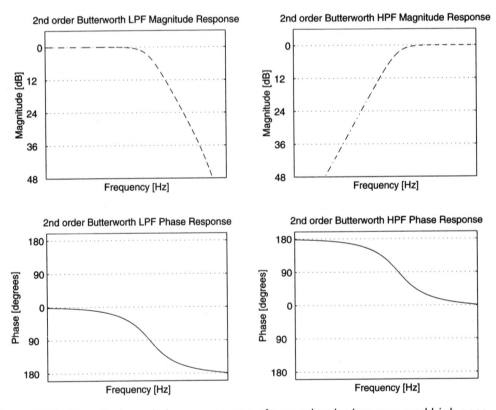

Figure 22-2. Magnitude and phase responses of second-order low-pass and high-pass Butterworth filters. The low-pass filter responses are on the left and the high-pass filter responses are on the right. Both filters have been designed for a crossover frequency of 1 kHz; the sampling rate is 44.1 kHz.

Digital Crossover

It is observed that the magnitude response is down −3 dB at the crossover frequency. The filters roll off at the rate of 12 dB/octave. Both of these are characteristics of second-order Butterworth filters. What happens when the output of these two sections is combined? *Figure 22-3* shows the individual responses and the combined response of the crossover design superimposed on the same plot.

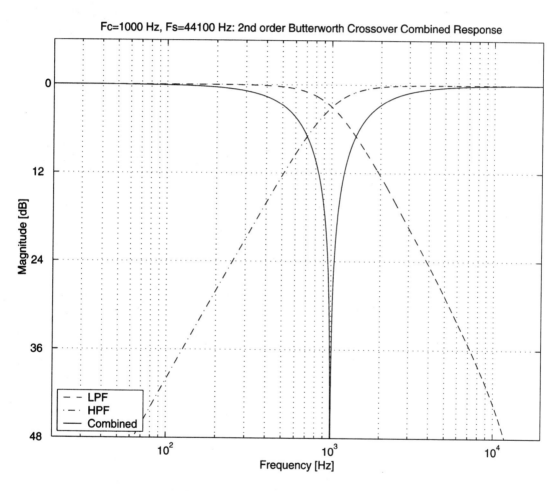

Figure 22-3. Individual and combined magnitude responses of second-order low-pass and high-pass Butterworth filters. Note the deep notch of the combined response at the crossover frequency.

DSP Filters

The result is a deep notch at the crossover frequency. This is clearly not usable for audio purposes. A cursory glance at the individual phase responses in *Figure 22-2* provides a clue. At the crossover frequency, for example, the low-pass filter has a phase of –90°, while the high-pass filter section has a phase of +90°. Combining these two sections results in signals that are 180° out of phase at the crossover frequency!

If the output of the one of the sections is inverted, then both sections end up being in phase at the crossover frequency. However, this results in a bump at the crossover frequency.

Fourth-order Butterworth filter

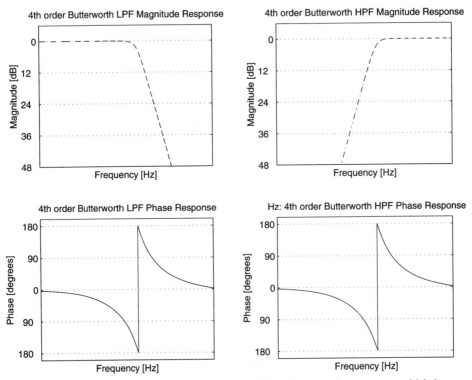

Figure 22-4. Magnitude and phase responses of fourth-order low-pass and high-pass Butterworth filters. The low-pass filter responses are on the left and the high-pass filter responses are on the right. Both filters have been designed for a crossover frequency of 1 kHz; the sampling rate is 44.1 kHz.

Digital Crossover

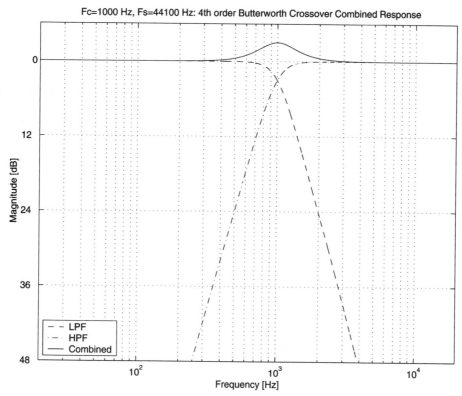

Figure 22-5. Individual and combined magnitude responses of fourth-order low-pass and high-pass Butterworth filters. Note the slight bump in the combined response at the crossover frequency.

Since the second-order Butterworth filter does not seem to be suitable for this purpose, one may examine the next even-order Butterworth filter — the fourth-order filter. *Figure 22-4* shows the magnitude and phase responses for the low-pass and high-pass sections.

It is observed that the magnitude response is down −3 dB at the crossover frequency. The filters roll off at a sharp rate of 24 dB/octave. Both of these are characteristics of fourth-order Butterworth filters. Further, the phase response in this case does not show the same relationship as the second-order sections. What happens when these two sections are combined? *Figure 22-5* shows the individual responses and

DSP Filters

the combined response of the crossover design superimposed on the same plot.

The result of combining the two sections is a bump of about 3 dB at the crossover frequency. This is a great improvement over the second-order crossover design; however, it would be worth investigating some other alternatives before one commits to using a fourth-order Butterworth filter for the design of a digital crossover.

Third-order Butterworth filter

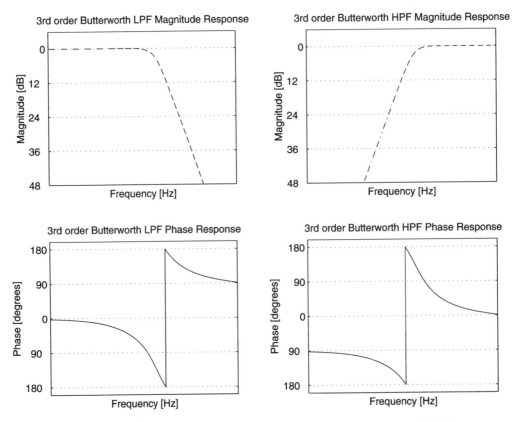

Figure 22-6. Magnitude and phase responses of third-order low-pass and high-pass Butterworth filters. The low-pass filter responses are on the left and the high-pass filter responses are on the right. Both filters have been designed for a crossover frequency of 1 kHz, the sampling rate is 44.1 kHz.

Digital Crossover

So far, the second- and fourth-order Butterworth filters (low-pass filter and high-pass filter) have been examined for the purpose of creating a digital crossover. One might consider an odd-order filter, namely a third-order Butterworth filter. (Odd-order filters are discussed in the Appendix.) *Figure 22-6* shows the magnitude and phase responses for the low-pass and high-pass sections of third-order Butterworth filters designed for a crossover frequency of 1 kHz.

It is observed that the magnitude response is down −3 dB at the crossover frequency. The filters roll off at an intermediate rate of 18 dB/octave. Both of these are characteristics of third-order Butterworth filters. The phase response in this case is quite interesting. If one takes into account the modulo range ±180° for phase, then the phase of the high-pass filter is observed to be exactly 90° behind the low-pass filter. What happens when these two sections are combined? *Figure 22-7* shows the individual responses and the combined response of the crossover design superimposed on the same plot.

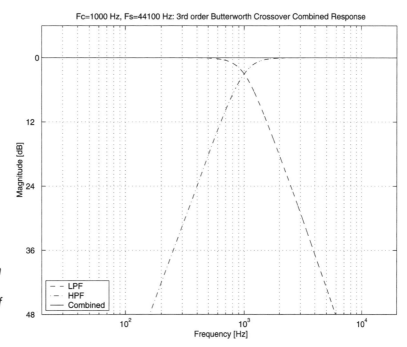

Figure 22-7. Individual and combined magnitude responses of third-order low-pass and high-pass Butterworth filters. Note that the combined response of the two filters is flat.

DSP Filters

The result of combining the two sections is a completely flat response. Finally, a filter that meets the criterion!

Fourth-order Linkwitz-Riley filter

S. Linkwitz and R. Riley were two Hewlett-Packard engineers who came up with the idea of cascading two identical Butterworth filter sections. The resulting filters are known as Linkwitz-Riley filters, and they always have an even order. Hence, a fourth-order Linkwitz-Riley filter consists of two identical second-order Butterworth filters that have been cascaded. Thus, a fourth-order Linkwitz-Riley filter is sometimes referred to as a second-order Butterworth squared. *Figure 22-8*

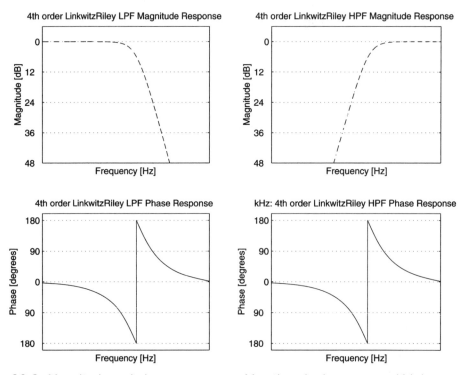

Figure 22-8. Magnitude and phase responses of fourth-order low-pass and high-pass Linkwitz-Riley filters. The low-pass filter responses are on the left and the high-pass filter responses are on the right. Both filters have been designed for a crossover frequency of 1 kHz; the sampling rate is 44.1 kHz.

Digital Crossover

shows the magnitude and phase responses for the low-pass and high-pass sections of fourth-order Linkwitz-Riley filters designed for a crossover frequency of 1 kHz.

It is observed that the magnitude response is down −6 dB at the crossover frequency (not −3 dB, like the Butterworth filters). This is because each of the underlying Butterworth filters reduces the magnitude response by 3 dB at the crossover frequency. The filters roll off at a rate of 24 dB/octave. Both of these are characteristics of fourth-order Linkwitz-Riley filters. The phase response in both sections is observed to be identical. What happens when these two sections are combined? *Figure 22-9* shows the individual responses and the combined response of the crossover design superimposed on the same plot.

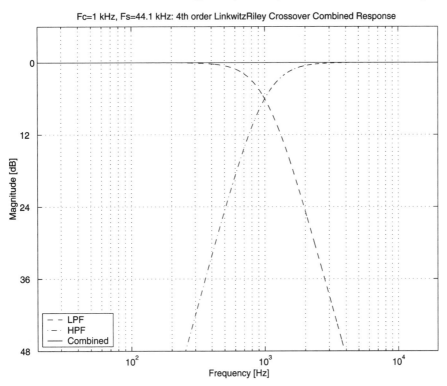

Figure 22-9. Individual and combined magnitude responses of fourth-order low-pass and high-pass Linkwitz-Riley filters. Note that the combined response of the two filters is flat.

DSP Filters

The result of combining the two sections is, once again, a completely flat response. Yet another filter that meets the criterion! The Linkwitz-Riley filter is beyond the scope of this book. In addition to the flat response shown above, this filter also has better off-axis cancellation properties as compared to other filters, especially on the vertical axis. This is increasingly the filter of choice in the professional audio arena.

Filter analysis revealed that the second- and fourth-order Butterworth filters are not suitable for crossover filter designs. The third-order Butterworth and the fourth-order Linkwitz-Riley filters met the criterion for crossover designs. In the remainder of this chapter we will focus on the third-order Butterworth filter design for the purpose of implementing a digital crossover.

Design Requirements

The goal of this design is to develop digital crossover filters such that a full bandwidth signal may be sent to loudspeakers with two drivers.

Specifications:
Channels: 2 (stereo)
Filters: Third-order low-pass filter to feed the Woofer
                        Third-order high-pass filter to feed the Tweeter
Crossover Frequency: Variable, but greater than 20 Hz and less than ¼ of the sampling rate
Sample Rate: Variable

This design will support stereo (2 channels) audio signals connected to two-way loudspeakers.

Filter Overview

A quick review of odd-ordered low-pass and high-pass filters may be helpful at this point. Equations for implementing these filters are

Digital Crossover

provided here for convenience along with the corresponding frequency and phase response curves. Details for these filters may be found in the Appendix.

Low-pass filter overview

$$H(z) = \frac{\alpha(1+z^{-1})}{1-\gamma z^{-1}} \cdot \prod_{k=1}^{M} \frac{\alpha_k(1+2z^{-1}+z^{-2})}{\frac{1}{2}-\gamma_k z^{-1}+\beta_k z^{-2}} \quad (22\text{-}1)$$

The third-order Butterworth low-pass filter is made up of two cascaded sections. The first section is a first-order section (as shown in the first line of Equation (22-1)) followed by a single second-order section (M = 1, on the second line of Equation (22-1)).

$$\gamma = \frac{\cos\theta_c}{1+\sin\theta_c} \qquad \alpha = (1-\gamma)/2 \quad (22\text{-}2a)$$

$$\beta_k = \frac{1}{2}\left(\frac{1-\frac{1}{2}d_k \sin\theta_c}{1+\frac{1}{2}d_k \sin\theta_c}\right) \quad \gamma_k = \left(\tfrac{1}{2}+\beta_k\right)\cos\theta_c \quad \alpha_k = \left(\tfrac{1}{2}+\beta_k-\gamma_k\right)/4 \quad (22\text{-}2b)$$

$$d_k = 2\sin\left(\frac{(2k-1)\pi}{4(M+\tfrac{1}{2})}\right) \quad (22\text{-}2c)$$

where, $\theta_c \equiv 2\pi f_c / f_s$ $k = 1...M$

Coefficient formulas for the odd-order low-pass Butterworth filter

Using Equations (22-2a) through (22-2c), it is possible to calculate the coefficients that will be required to implement the difference equations for the low-pass filter section of the crossover design. In order to do this, it is necessary to determine the crossover frequency, f_c, and the sampling rate of the system, f_s. Setting M = 1 will then yield the coefficients necessary (α, β, and γ).

DSP Filters

> **LOW-PASS:**
> $$y(n) = \alpha\,[x(n) + x(n-1)] + \gamma\,y(n-1)$$
> $$x_1(n) = y(n)$$
> $$y_1(n) = 2\{\alpha_1\,[x_1(n) + 2x_1(n-1) + x_1(n-2)] + \gamma_1\,y_1(n-1) - \beta_1\,y_1(n-2)\}$$
> $$x_2(n) = y_1(n) \qquad (22\text{-}3)$$
> $$\text{M}$$
> $$x_M(n) = y_{M-1}(n)$$
> $$y_M(n) = 2\{\alpha_M\,[x_M(n) + 2x_M(n-1) + x_M(n-2)] + \gamma_M\,y_M(n-1) - \beta_M\,y_M(n-2)\}$$

Difference equations to implement an odd-order low-pass Butterworth filter

The difference equations that are used to implement the digital filter are shown in Equation (22-3). The first line is the difference equation for the first-order low-pass filter section. Since a cascaded topology is being used, the output of the first-order low-pass filter section serves as an input to the second stage (which is a second-order low-pass filter section). This is implemented in the second line. Thereafter, alternate lines implement a second-order low-pass filter stage while the other line ensures a cascaded structure by treating the output of the previous stage as an input to the following stage.

The first line of Equation (22-3) describes the difference equation for a first-order low-pass filter, as shown in *Figure 4-3*. This may be implemented by summing the present input $x(n)$ with the past input $x(n-1)$. This result is multiplied by the coefficient *a* and to this is added the past output sample, $y(n-1)$ that has been scaled by γ. Details of this filter may be found in Chapter 4.

The second line of Equation (22-3) ensures that the output of the first-order low-pass filter, $y(n)$, becomes the input, $x_1(n)$, to the second-order low-pass filter in the second stage.

The third line of Equation (22-3) describes the difference equation for a second-order low-pass filter, as shown in *Figure 6-4*. This may be

Digital Crossover

implemented by summing the present and past inputs $x_1(n)$, $x_1(n-1)$ and $x_1(n-2)$ scaled respectively by a, $2a$, and a. To this is added to the past two outputs $y_1(n-1)$ and $y_1(n-2)$, which have respectively been scaled by γ and $-\beta$. This result is then multiplied by 2 to obtain the value of the present output, $y_1(n)$. Details of this filter are found in Chapter 6.

Since only a third-order low-pass filter is being implemented, the first four lines of Equation (22-3) are sufficient. The low-pass filter section would consist of two stages, a first-order stage cascaded to a second-order stage. Thus, $y_1(n)$ serves as the final output of low-pass filter section of the crossover design.

The magnitude and phase responses for a third-order Butterworth low-pass filter with different crossover frequencies are shown in *Figure 22-10*. The magnitude response is observed to be down by 3 dB at the crossover frequency, and the slope is observed to be 18 dB/octave, in keeping with the characteristics of a third-order Butterworth filter.

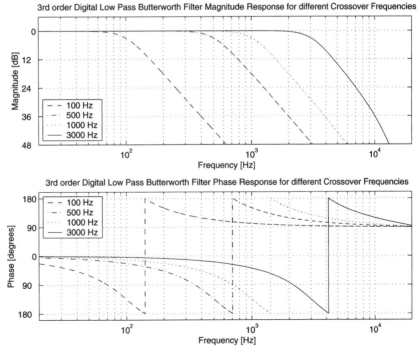

Figure 22-10. Magnitude and phase response of the third-order digital Butterworth low-pass filter using a sampling rate of 44.1 kHz: dashed line = 100 Hz; dot-dash line = 500 Hz; dotted line = 1000 Hz; solid line = 3000 Hz

DSP Filters

High-pass filter overview

$$H(z) = \frac{\alpha\left(1-z^{-1}\right)}{1-\gamma z^{-1}} \cdot$$

$$\prod_{k=1}^{M} \frac{\alpha_k\left(1-2z^{-1}+z^{-2}\right)}{\frac{1}{2}-\gamma_k z^{-1}+\beta_k z^{-2}} \quad (22\text{-}4)$$

The third-order Butterworth high-pass filter is made up of two cascaded sections. The first section is a first-order section (as shown in the first line of Equation 22-4) followed by a single second-order section (M = 1, on the second line of Equation (22-4)).

$$\gamma = \frac{\cos\theta_c}{1+\sin\theta_c} \qquad \alpha = (1+\gamma)/2 \qquad (22\text{-}5a)$$

$$\beta_k = \frac{1}{2}\left(\frac{1-\frac{1}{2}d_k \sin\theta_c}{1+\frac{1}{2}d_k \sin\theta_c}\right) \quad \gamma_k = (\tfrac{1}{2}+\beta_k)\cos\theta_c \quad \alpha_k = (\tfrac{1}{2}+\beta_k+\gamma_k)/4 \qquad (22\text{-}5b)$$

$$d_k = 2\sin\left(\frac{(2k-1)\pi}{4(M+\tfrac{1}{2})}\right) \qquad (22\text{-}5c)$$

where, $\theta_c \equiv 2\pi f_c / f_s$ $\qquad k = 1 \ldots M$

Coefficient formulas for the odd-order high-pass Butterworth filter

Using Equations (22-5a) through (22-5c), it is possible to calculate the coefficients that will be required to implement the difference equations for the high-pass filter section of the crossover design. Continuing with the example, it is assumed that the crossover frequency, f_c, and the sampling rate of the system, f_s, have been determined and that the value of M = 1 (for a third-order high-pass filter).

Digital Crossover

HIGH-PASS:
$$y(n) = \alpha\,[x(n) - x(n-1)] + \gamma\, y(n-1)$$
$$x_1(n) = y(n)$$
$$y_1(n) = 2\{\alpha_1\,[x_1(n) - 2x_1(n-1) + x_1(n-2)] + \gamma_1\, y_1(n-1) - \beta_1\, y_1(n-2)\}$$
$$x_2(n) = y_1(n)$$
$$\mathbf{M}$$
$$x_M(n) = y_{M-1}(n)$$
$$y_M(n) = 2\{\alpha_M\,[x_M(n) - 2x_M(n-1) + x_M(n-2)] + \gamma_M\, y_M(n-1) - \beta_M\, y_M(n-2)\} \qquad (22\text{-}6)$$

Difference equations to implement an odd-order high-pass Butterworth filter

The difference equations that are used to implement the digital filter are shown in Equation (22-6). The first line is the difference equation for the first-order high-pass filter section. Since a cascaded topology is being used, the output of the first-order high-pass filter section serves as an input to the second stage (which is a second-order high-pass filter section). This is implemented in the second line. Thereafter, alternate lines implement a second-order high-pass filter stage while the other line ensures a cascaded structure by treating the output of the previous stage as an input to the following stage.

The first line of Equation (22-6) describes the difference equation for a first-order high-pass filter, as shown in *Figure 5-3*. This may be implemented by subtracting the past input $x(n-1)$, from the present input $x(n)$. This result is multiplied by the coefficient *a* and to this is added the past output sample, $y(n-1)$ that has been scaled by γ. Details of this filter may be found in Chapter 5.

The second line of Equation (22-6) ensures that the output of the first-order high-pass filter, $y(n)$, becomes the input, $x_1(n)$, to the second-order high-pass filter in the second stage.

The third line of Equation (22-6) describes the difference equation for a second-order high-pass filter, as shown in *Figure 7-4*. This may be

DSP Filters

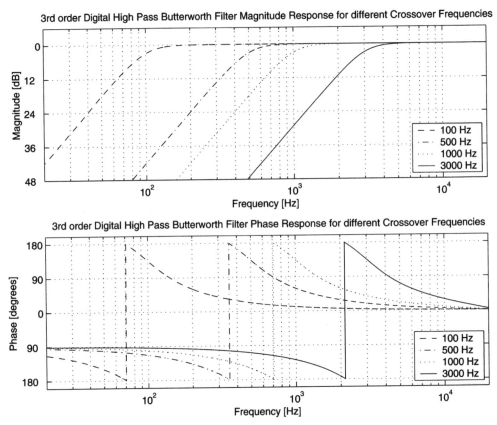

Figure 22-11. Magnitude and phase response of the third-order digital Butterworth high-pass filter using a sampling rate of 44.1 kHz: dashed line = 100 Hz; dot-dash line = 500 Hz; dotted line = 1000 Hz; solid line = 3000 Hz

implemented by summing the present and past inputs $x_1(n)$, $x_1(n-1)$ and $x_1(n-2)$ scaled respectively by a, $-2a$, and a. To this is added the past two outputs $y_1(n-1)$ and $y_1(n-2)$, which have respectively been scaled by γ and $-\beta$. This result is then multiplied by 2 to obtain the value of the present output, $y_1(n)$. Details of this filter are found in Chapter 7.

Since only a third-order high-pass filter is being implemented, the first four lines of Equation (22-6) are sufficient. The high-pass filter section would then consist of two cascaded stages — a first-order stage

Digital Crossover

followed by a second-order stage. Thus, $y_1(n)$ serves as the final output of high-pass filter section of the crossover design.

The magnitude and phase responses for third-order Butterworth high-pass filter with different crossover frequencies are shown in *Figure 22-11*. The magnitude response is observed to be down by 3 dB at the crossover frequency, and the slope is observed to be 18 dB/octave in keeping with the characteristics of a third-order Butterworth filters.

Functional Blocks

Conceptually, the crossover design consists of three basic components: the control section, the coefficient calculation section, and the filter section. The control section of the crossover design allows the user to specify the crossover frequency f_c, which is applied to both the low-pass filter and high-pass filter sections. The function of the coefficient calculation blocks is to calculate the coefficients — α and γ for the first-order filter sections and α, β, and γ for the second-order filter sections. The filter section utilizes the different coefficients that have been calculated for each of the filter sections, and processes the incoming audio signal. The implementation of these components is outlined below.

Control section

The control section of the crossover design can be implemented in either hardware, for example, in an electronic crossover used for live sound applications, or in software for PC applications. Regardless of the type of design required (hardware or software), the control section must provide a numerical representation of the control knob position that is being dialed in by the user to select the crossover frequency. In a hardware design, an analog potentiometer may be used to feed an A/D converter used for control data. The digital output of the A/D converter may be used directly in the calculation of the filter coefficients. In a software application, such as in a PC, the control knob positions as displayed in a

DSP Filters

graphical user interface (GUI) can be read by a software module and applied directly, or even greater frequency resolution may be obtained by allowing the user to type in the desired crossover frequency.

Coefficient calculation section

The crossover frequency value obtained from the previous section may be used to compute the coefficients required in the difference equations. These have been summarized in Equation (22-2) and Equation (22-5). In this example, we have chosen to apply a fixed sample rate (44.1 kHz).

Filter section

Since third-order Butterworth filters are being implemented for the crossover design, the same input signal is processed by a third-order low-pass filter as well as a third-order high-pass filter. Thus, a single audio input yields two outputs, resulting in a two-way crossover design. The output of the low-pass filter section will then be amplified and directed to the driver handling the lower frequencies, namely the woofer. The output of the high-pass filter section, on the other hand, will also be amplified but directed to the driver handling the higher frequencies, namely the tweeter.

Each third-order filter comprises of a first-order section cascaded with a second-order section. The input signal is always fed to both the first-order low-pass filter *and* the first-order high-pass filter. The output of the first-order low-pass filter is then treated as an input to the second-order low-pass filter. Similarly, the output of the first-order high-pass filter is treated as an input to the second-order high-pass filter.

This is how the crossover design takes a single input signal and splits it into two frequency components (low and high) by applying two third-order Butterworth filters (low-pass filter and high-pass filter) to it.

Control Flow Descriptions

The coefficient calculation section utilizes two routines to determine the coefficients α, β, and γ. When the crossover frequency control is adjusted in the control section, the routine *calc_coef_for_new_f0(double dv, int i)* is activated to update the new crossover frequency value in the matrix table. Note that each of the filter sections for a crossover design is tuned to the same frequency. This is different from the previous cascaded examples. Hence, there is really no need to pass an index (*int i*) as a variable (as was necessary in previous examples). However, we shall continue to do so to be consistent. Any adjustment of the crossover frequency will activate the *calc_coef_for_new_gain(double dv, int i)* routine, which will result in the recalculation of new coefficients for the corresponding filter. Once again, there is no need to pass the index, since all coefficients must be recalculated any time there is a change in the crossover frequency.

Updating the sample rate

When the sample rate of the input data changes, the *calc_coeff_for_new_fs(double dv)* routine is called. This routine receives only one variable, *dv*, containing the new sample rate in Hertz. The sample rate change affects all filters and therefore no index, *i*, is required for this operation. The sample rate affects the value of the normalized crossover frequency (θ) for each filter section which is calculated in the *calc_coef_for_new_f0(double dv , int i)* routine. Thus, the *calc_coef_for_new_fs(double dv)* routine updates the sample rate values in the matrix and then activates the *calc_coef_for_new_f0(double dv, int i)* routine to recalculate the new θ for each filter section. The routine *calc_coef_for_new_gain(double dv, int i)* is then called to complete the coefficient calculation process given the updated θ value for each filter section in a loop.

DSP Filters

Updating the crossover frequency

A change in either sample rate or crossover frequency activates the routine *calc_coeff_for_new_f0(double dv, int i)*. Two variables are passed to this routine: the crossover frequency *dv* and the redundant index *i*. The updated crossover frequency is tested to verify that it falls between the system-defined limits (*min_f0, max_f0*). The new crossover frequency is then copied into each individual filter. Finally, the routine *calc_coeff_for_new_gain(double dv, int i)* is called where θ is calculated as a function of the crossover frequency and the sample rate to calculate the new coefficients based on the new θ value for each filter section in a loop.

Updating the coefficients

The *calc_coeff_for_new_gain(double dv, int i)* routine is where the coefficients are calculated based on changes in sample rate or crossover frequency. Two variables are passed to this routine — the gain (*dv*) and the index *i*. The coefficients are to be determined based on the type of filter to be implemented. A first-order low-pass filter is implemented when the index value is 0, a second-order low-pass filter is implemented when the index is 1, a first-order high-pass filter is implemented when the index is 2, and a second-order high-pass filter is implemented when the index is 3. Since all four filters are affected by any change to the crossover frequency or the sampling rate, no index is passed into this function. Instead, the function internally cycles through the range of index values.

Hence, when this routine is called, it calculates the coefficients in each of the four filter sections by using Equation (22-2) and Equation (22-5). Once the coefficients α, β, and γ have been determined, they are stored for use by the filter corresponding to the index *i*.

Digital Crossover

Software Description

All filter-related information is stored in a two-dimensional array or matrix as shown in *Table 22-1*. The matrix is used to store variables required for processing each of the filters. Each column contains the variables associated with one of the four filters in the filter section and is referenced in the following routines by the value *i*. Each row represents a particular variable that is required for processing each of the filters.

Table 22-1: Cascaded crossover filter variable array: stages[i]->variable

Variable	*i* -> 0 First order LPF	1 Second Order LPF	2 First Order HPF	3 Second Order HPF
min_f0 - Minimum Crossover Filter Frequency	20.0	20.0	20.0	20.0
f0 - Crossover Filter Frequency	1000.0	1000.0	1000.0	1000.0
max_f0 - Maximum Crossover Filter Frequency	11025.0	11025.0	11025.0	11025.0
min_gain - Minimum Crossover Filter gain	-20	-20	-20	-20
Gain - Crossover Filter gain	0	0	0	0
max_gain - Maximum Crossover Filter gain	20	20	20	20
min_Q - Minimum Crossover Filter Q factor	0.1	0.1	0.1	0.1
Q - Crossover Filter Q factor	1.0	1.0	1.0	1.0
max_Q - Maximum Crossover Q factor	20	20	20	20
x - Crossover Filter Input Sample	0	0	0	0
x1 - Previous Crossover Filter Input Sample	0	0	0	0
x2 – Previous, Previous Crossover Filter Input Sample	-	0	-	0
y - Crossover Filter Output Sample	0	0	0	0
y1 - Previous Crossover Filter Output Sample	0	0	0	0
y2 – Previous, Previous Crossover Filter Output Sample	-	0	-	0
theta0 - $2*pi*f_0/f_s$	.1425	.1425	.1425	.1425
beta - Crossover Filter feedback coefficient	-	.4337	-	.4337
gamma - Crossover Filter feedback coefficient	.8668	.9242	.8668	.9242
alpha - Crossover Filter feed-forward coefficient	.9334	.0024	.9334	.4645

The following software routines implement the crossover filter design. The routines are all provided as software modules written in C++.

The implementation begins with the initialization routine required to initialize the variable array to a known stable state prior to activation of the crossover. Initialization is performed only once prior to activation or following any system reset condition.

DSP Filters

Initializing the variable matrix

The initialization routine establishes several parameters necessary to allow the crossover design to meet the goals defined at the outset of this section. A detailed analysis of the initialization routine is provided.

```
// Implementation of a Cascaded Crossover Filter Block
1       const double TWO_PI = 2.0 *3.14159265358979323846;
2       const double PI = 3.14159265358979323846;
3       const int NUM_BANDS = 4;

4       CCascaded_Crossover::CCascaded_Crossover()
5       {
6       int i;
7       min_beta = -0.4999;
8       max_beta = 0.4999;
9       stages = new (CCookFilterStage * [NUM_BANDS]);
10      stages[0] = new Clowpass1FilterStage;
11      stages[NUM_BANDS/2] = new Chighpass1FilterStage;

12      for (i = 1; i < NUM_BANDS/2; i++)
13      {
14      stages[i] = new CLowpassFilterStage;
15      stages[i+NUM_BANDS/2] = new CHighpassFilterStage;
16      }

17      for (i = 0; i < NUM_BANDS; i++)
18      {
19      stages[i]->min_f0 = 20.0;
20      stages[i]->min_gain = 0.;
21      stages[i]->max_gain = 5.0;
22      stages[i]->min_Q = 0.1;
23      stages[i]->max_Q = 20.0;
24      stages[i]->gain = 1.0;
25      stages[i]->Q = sqrt(2.0) * 3.0;
26      stages[i]->f0 = 1000;
27      stages[i]->max_f0 = 11025.0;
28      }
29      num_filter_stages = NUM_BANDS;
30      calc_coeff_for_new_f0(1000.0,0);
31      enabled = 1;
32      num_outputs = 2;
33      }
```

Digital Crossover

- Lines 1, 2, and 3 define constants "2pi" ($2*\pi$) and "pi" (π) as well as the number of filter sections (4 – first-order low-pass filter, second-order low-pass filter, first-order high-pass filter and second-order high-pass filter).

- Lines 4, 5, and 33 define the boundaries of this routine. Everything between these lines constitutes the main initialization routine.

- Line 6 initializes a variable, i.

- Lines 7 and 8 set up maximum and minimum values for the variable *beta*.

- Line 9 calls a routine that clears the memory used for the variable matrix.

- Lines 10 and 11 invoke constructors for the first-order low-pass filter and high-pass filter, respectively. Thus, stages[0] will be a first-order low-pass filter, while stages[NUM_BANDS/2] will be a first-order high-pass filter.

- Lines 12 through 16 create a loop within which constructors are invoked for second-order low-pass filter and high-pass filter sections. At the end of this, half the filter sections will be low-pass and the other half will be high-pass. Further, other than a single first-order section for low-pass filter and high-pass filter, the remaining filter sections are second-order. Also, the constructors establish memory in the system for the matrix variables and associate the variables with the rows of the matrix.

- Lines 19 through 27 initialize the variables in the matrix for each filter, as shown.

    Line 19 initializes the minimum crossover frequency for the filter.
    Line 20 initializes the minimum gain for the filter.
    Line 21 initializes the maximum gain for the filter.
    Line 22 initializes the minimum Q for the filter.

DSP Filters

> Line 23 initializes the maximum Q for the filter.
>
> Line 24 initializes the gain for the filter.
>
> Line 25 initializes the Q for the filter.
>
> Line 26 initializes the crossover frequency for the filter.
>
> Line 27 initializes the maximum crossover frequency for the filter.
>
> Only min_f0, f0, and max_f0 are actually controlled by the filter. The remaining variables are a required overhead in the DirectX control interface

- Line 29 establishes num_filter_stages equal to NUM_BANDS, which in this case is 4.
- Line 30 invokes the function *calc_coeff_for_new_f0()* with the parameter 1000 Hz. This ensures that the crossover frequency is set to 1000 Hz, further, as shown below, this function will, in turn, call the function *calc_coeff_for_new_gain()*, which will calculate the coefficients for each filter.
- Line 31 initializes the variable, *enabled*, which indicates that initialization has been completed and the crossover filter can now be enabled.
- Line 32 intializes the variable, *num_outputs*, to 2 to denote that there are two outputs (one for the low-pass filter section and the other for the high-pass filter section).

The four constructors that are used during system initialization to clear the variables indicated for each column in the matrix are as follows:

First-order low-pass filter *Clowpass1FilterStage* constructor (details in Chapter 4):

```
Clowpass1FilterStage::Clowpass1FilterStage()
{
        x1 = 0;
        y1 = 0;
}
```

Digital Crossover

First-order high-pass filter *Chighpass1FilterStage* constructor (details in Chapter 5):

```
Chighpass1FilterStage::Chighpass1FilterStage()
{
        x1 = 0;
        y1 = 0;
}
```

Second-order low-pass filter *CLowpassFilterStage* constructor (details in Chapter 6):

```
CLowpassFilterStage::CLowpassFilterStage()
{
        x1 = 0;
        x2 = 0;
        y1 = 0;
        y2 = 0;
}
```

Second-order high-pass filter *CHighpassFilterStage* constructor (details in Chapter 7):

```
CHighpassFilterStage::CHighpassFilterStage()
{
        x1 = 0;
        x2 = 0;
        y1 = 0;
        y2 = 0;
}
```

Calculating the coefficients

The heart of the processing boils down to implementing the underlying filters. In turn, the nature of the filter is determined by the coefficients. Any time there is a change in either the crossover frequency or the sampling rate, the filter coefficients need to be recalculated. The routine *calc_coeff_for_new_gain()* below performs these computations.

DSP Filters

```
1      void CCascaded_Crossover::calc_coeff_for_new_gain (double dv, int i)
2      {
3      double dk;
4      int k;
5      int M;
6      CCookFilterStage *st;
7      M = num_filter_stages/2;

8      stages[0]->gain = dv;
9      stages[M]->gain = dv;

10     // Cascaded Lowpass
11     for(k=0; k < M; k++)
12     {
13     st = stages[k];
14     st->theta0 = 2.0 * PI * st->f0 / fs;

15     if (k == 0)   // first order low-pass filter section
16     {
17     st->gamma = cos(st->theta0) /
(1   + sin(st->theta0));
18     st->alpha = (1 - st->gamma) / 2.0;
19     }
20     else          // second order low-pass filter sections
21     {
22     dk = 2.0 * sin((((2.0 * k)) - 1.0) * PI)/(4.0 * (M + 0.5));
23     st->beta = 0.5 * (1.0 - (dk/2.0)* sin(st->theta0)) /
(1    + (dk/2.0)* sin(st->theta0));
24     st->gamma = (0.5 + st->beta) * cos(st->theta0);
25     st->alpha = (0.5 + st->beta - st->gamma)/ 4.0;
26     }
27     }

28     // Cascaded Highpass
29     for(k=0; k < M; k++)
30     {
31     st = stages[M + k];
32     st->theta0 = 2.0 * PI * st->f0 / fs;

33     if (k == 0)   // first order high-pass filter section
34     {
35     st->gamma = cos(st->theta0) /
(1    + sin(st->theta0));
36     st->alpha = (1 + st->gamma) / 2.0;
```

Digital Crossover

```
37      }
38      else
39      {
40      dk = 2.0 * sin((((2.0 * k) - 1.0) * PI)/(4.0 * (M + 0.5)));
41      st->beta = 0.5 * (1.0 - (dk/2.0)* sin(st->theta0)) /
(1      + (dk/2.0)* sin(st->theta0));
42      st->gamma = (0.5 + st->beta) * cos(st->theta0);
43      st->alpha = (0.5 + st->beta + st->gamma)/ 4.0;
44      }
45      }
46      }
```

- Lines 1, 2, and 46 define the boundaries for this routine.

- Lines 3-7 are used to define different variables.

- Lines 8 and 9 are where the new gain passed through the variable, *dv*, are incorporated into the first-order low-pass filter and high-pass filter sections.

- Lines 10, 11, 12, and 27 set up the boundaries for the code where coefficients are calculated for each of the low-pass filter sections.

- Line 13 establishes *st* as a pointer to the variables in column *k*, note that this only runs through half the values of the variable matrix, namely the low-pass filter sections.

- Line 14 calculates the new normalized crossover frequency, *theta0*.

- Lines 15 through 19 are only executed for *k* equal to zero, that is for the first-order low-pass filter. The coefficients α and γ are calculated here.

- Lines 20 through 26 are only executed for *k* not equal to zero, that is for the second-order cascaded low-pass filter sections. The coefficients α, β, and γ are calculated here.

- Lines 28 through 45 follow the exact same logic as above, but for the high-pass filter sections. There is a one-to-one correspondence between lines 10 through 27 and lines 28 through 45.

DSP Filters

Updating the crossover frequency

When a new crossover frequency is input, it must be verified that it lies within established bounds. Each filter is made aware of the new crossover frequency. Further, this requires that the filter coefficients be recalculated. The routine that does these tasks, *calc_coeff_for_new_f0()*, is described below.

```
1     void CCascaded_Crossover::calc_coeff_for_new_f0 (double dv, int i)
2     {
3     if (dv > stages[0]->max_f0)      dv = stages[0]->max_f0;
4     else if (dv < stages[0]->min_f0) dv = stages[0]->min_f0;
5     for (i = 0; i < NUM_BANDS; i++)
6     {
7     stages[i]->f0 = dv;
8     }
9     calc_coeff_for_new_gain (stages[0]->gain, 0);
10    }
```

- Lines 1, 2, and 10 define the boundaries of this routine. The routine receives two variables: crossover frequency *(dv)*, and filter section index *i*.

- Line 3 tests the new f_0 verifying that it does not exceed the maximum crossover frequency boundary *(max_f0)*.

- Line 4 tests the new f_0 verifying that it is not lower than the minimum crossover frequency boundary *(min_f0)*.

- Lines 5 through 8 are used to set up a loop to copy the new crossover frequency value *dv* in the f_0 location in the column indicated by *i*.

- Line 9 calls the routine *calc_coeff_for_new_gain (stages[0]->gain,0)*. This routine calculates the new coefficients α, β, and γ given the updated crossover frequency.

Digital Crossover

Updating the sampling rate

A change in the sampling rate affects a couple of things:

1. It directly affects the upper bound for allowable crossover frequency values.

2. It ends up altering the "normalized crossover frequency," which in turn requires the filter coefficients to be recalculated.

This chain of events is carried out by the routine, *calc_coeff_for_new_fs()*, described below.

```
1    void CCascaded_Crossover::calc_coeff_for_new_fs (double dv)
2    {
3    int i;
4    fs = dv;
5    for (i = 0; i < NUM_BANDS; i++)
6    {
7    stages[i]->max_f0 = fs/4.0;
8    }
9    calc_coeff_for_new_f0 (stages[0]->f0, 0);
10   }
```

- Lines 1, 2, and 10 define the boundaries of this routine. The routine receives one variable: new sample rate. The new sample rate is denoted by the variable name *dv*.

- Line 3 is used to initialize a variable.

- Line 4 saves the new sample rate *dv* as the variable f_s.

- Lines 5 through 8 sets up a loop within which the maximum allowable value of the crossover frequency is recalculated, based on the new sampling rate for each filter.

- Line 9 calls the routine *calc_coeff_for_new_f0 (stages[0]->f0,0)*. This function has been described above. It updates additional variables in the matrix related to sample rate and in turn also calls *calc_coeff_for_new_gain()* that results in the filter coefficients being recalculated.

DSP Filters

Implementing the filters

The main filter loop is implemented with the following *execute_filter_block_in_place(double *in, int output_index)* routine. The input sample is passed to the routine via the pointer **in* and is initially stored as the *x* variable for the first-order filter (index $i = 0$). This may either be a low-pass filter or high-pass filter section depending on the value of the variable *output_index*. If *output_index* is 0, then the signal is passed through all the cascaded low-pass filter sections (including the first-order section). When the *output_index* has a non-zero value then the high-pass filter sections are invoked. The routine *execute_filter_stage()* is called to process the input sample with the appropriate low-pass filter or high-pass filter section. Processing takes place such that the output of one section is passed as the input to the following cascaded section. When all the filter stages have processed the signal, the final output is written back into the pointer **in*, thus achieving in-place processing.

```
1      void CCascaded_Crossover::execute_filter_block_in_place(double
*in, int output_index)
2      {
3          int i;
4          double input = *in;

5          // Execute only the Cascaded Lowpass filter for output 0
6          if (output_index == 0)
7          {
8              stages[0]->x = input;
9              stages[0]->execute_filter_stage();

10             for (i= 1; i< (NUM_BANDS/2); i++)
11             {
12                 stages[i]->x = stages[i-1]->y;
13                 stages[i]->execute_filter_stage();
14             }

15             *in = stages[(NUM_BANDS/2)-1]->y;
16         }
17         // Execute only the Cascaded Highpass filter for output 1
18         else
```

```
19      {
20          int j = NUM_BANDS/2;
21          stages[j]->x = input;
22          stages[j]->execute_filter_stage();

23          for (i= j+1; i<NUM_BANDS; i++)
24          {
25              stages[i]->x = stages[i-1]->y;
26              stages[i]->execute_filter_stage();
27          }

28          *in = stages[NUM_BANDS-1]->y;
29      }
30  }
```

- Lines 1, 2, and 30 define the boundaries of this routine.

- Line 3 defines the variable *i* as an integer that is used to reference variables in a particular column in the matrix shown in *Table 22-1*. There are four filters in the table; however, only half the filters are low-pass filters and the other half are high-pass filters. Since this routine is only called on to process either low-pass filter or high-pass filter sections, *i* will take on the values of 0 or 1. The value 0 would correspond to a first-order section and a non-zero value would correspond to second-order sections. Each iteration is referred to as a section of the crossover filter. Each section implements one filter and has a dedicated column in the matrix allocated for its use.

- Line 4 defines a variable input initialized to the pointer value *in*. *in* is a pointer to the input sample.

- Lines 6 through 16 are only executed if the variable *output_index* is 0 (that is, process low-pass filter sections)

    Line 8 stores the input sample value in the first filter's variable x in the first column of the matrix.

    Line 9 calls the routine execute_filter_stage( ), *which processes the first filter (first-order low-pass filter).*

DSP Filters

> Lines 10, 11, and 14 are the boundaries for an iterative processing loop for all the other second-order cascaded low-pass filter sections
>
> Line 12 ensures that the output of the previous low-pass filter section becomes the input to the present low-pass filter section.
>
> Line 13 calls the routine execute_filter_stage( ), which processes the subsequent filters (second-order low-pass filter)
>
> Line 15 passes the output of the last low-pass filter section to the pointer *in, this is the final output of the cascaded low-pass filter sections of the crossover filter. *in is thus pointing to the output samples that will be directed to the woofer.

- Lines 18 through 29 follow the exact same logic as above, but for the high-pass filter sections. There is a one-to-one correspondence between lines 6 through 16 and lines 18 through 29 with the addition of line 20. A variable, *j*, has been introduced to make the indexing easier in the high-pass filter section of the code. Similar to line 15, line 28 passes the output of the last high-pass filter section to the pointer *in*. These output samples will ultimately be directed to the tweeter.

What follows is a brief description of the code that actually implements the filtering operation for each of the underlying filters that is used in the cascaded crossover filter design. The four filters are: first-order low-pass filter, first-order high-pass filter, second-order low-pass filter, and second-order high-pass filter. The variables and signal processing associated with each of these filters is described below.

The *Clowpass1FilterStage::execute_filter_stage()* implements the first-order low-pass filter. The low-pass filter coefficients and state variables are stored in the first column of the matrix. The variables utilized in this routine are listed below:

```
x  - Input sample
x1 - past input sample
y  - Output sample
y1 - past output sample
```

Digital Crossover

This routine implements the difference equation for the first-order low-pass filter as shown in the first line of Equation (22-3).

```
1    void Clowpass1FilterStage::execute_filter_stage()
2    {
3        y = alpha * (x + x1) + gamma * y1 ;
4        x1 = x;
5        y1 = y;
6    }
```

- Lines 1, 2, and 6 define the boundaries of this routine.
- Line 3 implements the first-order low-pass filter difference equation for the corresponding filter.
- Lines 4 and 5 update the filter state variables (x1, and y1) for the corresponding filter.

The *Chighpass1FilterStage::execute_filter_stage()* implements the first-order high-pass filter. The high-pass filter coefficients and state variables are stored in the third column of the matrix. The variables utilized in this routine are listed below:

```
x  - Input sample
x1 - past input sample
y  - Output sample
y1 - past output sample
```

This routine implements the difference equation for the first-order high-pass filter as shown in the first line of Equation (22-6).

```
1    void Chighpass1FilterStage::execute_filter_stage()
2    {
3        y = alpha * (x - x1) + gamma * y1 ;
4        x1 = x;
5        y1 = y;
6    }
```

DSP Filters

- Lines 1, 2, and 6 define the boundaries of this routine.
- Line 3 implements the first-order high-pass filter difference equation for the corresponding filter.
- Lines 4 and 5 update the filter state variables (x1, and y1) for the corresponding filter.

The *CLowpassFilterStage::execute_filter_stage()* implements the second-order low-pass filter. The low-pass filter coefficients and state variables are stored in the second column of the matrix. The variables utilized in this routine are listed below:

```
x  - Input sample
x1, x2 - past input samples
y  - Output sample
y1, y2 - past output samples
alpha, beta, gamma - filter coefficients
```

This routine implements the difference equation for the second-order low-pass filter as shown in the third line of Equation (22-3).

```
1       void CLowpassFilterStage::execute_filter_stage()
2       {
3           y = 2*((alpha * (x + 2.0 * x1+ x2) + gamma * y1 - beta * y2);
4           x2 = x1;
5           x1 = x;
6           y2 = y1;
7           y1 = y;
8       }
```

- Lines 1, 2, and 8 define the boundaries of this routine.
- Line 3 implements the second-order low-pass filter difference equation for the corresponding filter.
- Lines 4 through 7 update the filter state variables (x1, x2, y1, and y2) for the corresponding filter.

Digital Crossover

The *CHighpassFilterStage::execute_filter_stage()* implements the second-order high-pass filter. The high-pass filter coefficients and state variables are stored in the fourth column of the matrix. The variables utilized in this routine are listed below:

```
x - Input sample
x1, x2 - past input samples
y - Output sample
y1, y2 - past output samples
alpha, beta, gamma - filter coefficients
```

This routine implements the difference equation for the second-order high-pass filter as shown in the third line of Equation (22-6).

```
9      void CHighpassFilterStage::execute_filter_stage()
10     {
11     y = 2*((alpha * (x - 2.0 * x1+ x2) + gamma * y1 - beta * y2);
12     x2 = x1;
13     x1 = x;
14     y2 = y1;
15     y1 = y;
16     }
```

- Lines 1, 2, and 8 define the boundaries of this routine.

- Line 3 implements the second-order high-pass filter difference equation for the corresponding filter.

- Lines 4 through 7 update the filter state variables (x1, x2, y1, and y2) for the corresponding filter.

Appendix

Odd-Order Filters

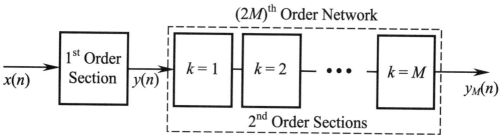

Figure A-1. Odd-order cascaded filter consisting of first-order section followed by an even-order cascaded filter made up of M second-order sections. Total order of filter: N = 2M+1

The reader may have observed that with the exception of the first-order low-pass and high-pass filters described in Chapters 4 and 5, all other filters discussed in this text have been *even-order* filters. In the same way that an odd number can be generated by adding the number *1* to any even number, an *odd-order* filter can be constructed by simply cascading a first-order section with an even-order network, such as those discussed in Chapters 12 and 13.

DSP Filters

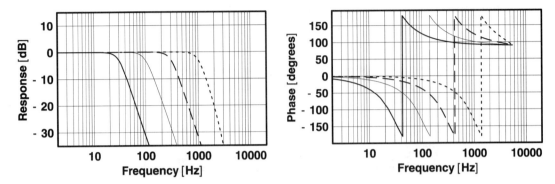

Figure A-2. Gain and phase response of digital third-order (N=3) low-pass Butterworth filter with $f_s = 11025$: solid line, $f_c = 30$; thin solid line, $f_c = 100$; dashed line, $f_c = 300$; and dotted line, $f_c = 1000$

It should be noted that since there is no first-order bandpass or bandstop filter, odd-order bandpass and band-stop filters are not created in this way. Therefore, we will focus our attention on the odd-order low-pass and high-pass *maximally flat* Butterworth filters.

Digital Low-Pass Filter

The odd-order low-pass frequency response function can be expressed by combining the first-order function from Chapter 4, Equation (4-7), and the cascaded function from Chapter 12, Equation (12-12):

$$H(z) = \frac{\alpha\left(1+z^{-1}\right)}{1-\gamma z^{-1}} \cdot \prod_{k=1}^{M} \frac{\alpha_k\left(1+2z^{-1}+z^{-2}\right)}{\frac{1}{2}-\gamma_k z^{-1}+\beta_k z^{-2}}$$

(A-1)

where the order is $N = 2M+1$. The coefficients in Equation (A-1) are described by Equations (4-8) for the first-order section, and by Equations (12-10) and (12-11) for the cascaded even-order section.

Appendix: Odd-Order Filters

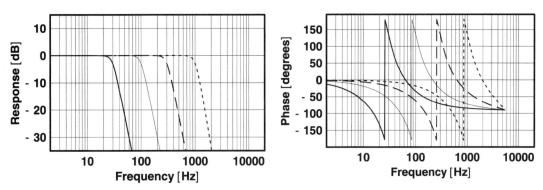

Figure A-3. Gain and phase response of digital fifth-order (N=5) low-pass Butterworth filter with $f_s = 11025$: solid line, $f_c = 30$; thin solid line, $f_c = 100$; dashed line, $f_c = 300$; and dotted line, $f_c = 1000$

There are two additional requirements that must be met in order to implement the odd-order filter:

1. The center frequency of the first-order section is set equal to the center frequency of the cascaded network.
2. The variable M in the damping factor formula of Equation (12-11) is replaced with M+1/2, so that $M = (N-1)/2$ where N is the total filter order.

The first-order and cascaded coefficient formulas are repeated, with the above requirements in Equations (A-2a), (A-2b), and (A-2c).

$$\gamma = \frac{\cos\theta_c}{1+\sin\theta_c} \qquad \alpha = (1-\gamma)/2 \qquad \text{(A-2a)}$$

$$\beta_k = \frac{1}{2}\left(\frac{1-\frac{1}{2}d_k\sin\theta_c}{1+\frac{1}{2}d_k\sin\theta_c}\right) \qquad \gamma_k = \left(\tfrac{1}{2}+\beta_k\right)\cos\theta_c \qquad \alpha_k = \left(\tfrac{1}{2}+\beta_k-\gamma_k\right)/4 \qquad \text{(A-2b)}$$

$$d_k = 2\sin\left(\frac{(2k-1)\pi}{4(M+\tfrac{1}{2})}\right) \qquad \text{(A-2c)}$$

$$\text{where,} \quad \theta_c \equiv 2\pi f_c/f_s \qquad k = 1\ldots M$$

Coefficient formulas for the odd-order low-pass Butterworth filter

DSP Filters

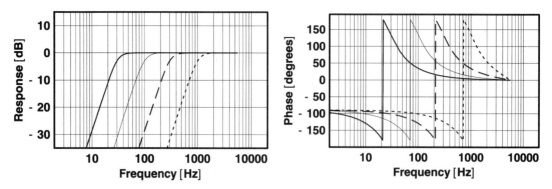

Figure A-4. Gain and phase response of digital third-order (N=3) high-pass Butterworth filter with $f_s = 11025$: solid line, $f_c = 30$; thin solid line, $f_c = 100$; dashed line, $f_c = 300$; and dotted line, $f_c = 1000$

Digital High-Pass Filter

As we have seen in previous chapters, the differences between the low-pass and high-pass filters amount to nothing more than a few sign changes in the response function and coefficient formulas. The transfer function for the high-pass can be written by referring back to the first-order function described by Equation (5-7) and the cascaded function of Equation (13-12):

$$H(z) = \frac{\alpha \left(1 - z^{-1}\right)}{1 - \gamma z^{-1}} \cdot \prod_{k=1}^{M} \frac{\alpha_k \left(1 - 2 z^{-1} + z^{-2}\right)}{\frac{1}{2} - \gamma_k z^{-1} + \beta_k z^{-2}} \quad \text{(A-3)}$$

As in the previous low-pass case, the range of k is from 1 to $M = (N-1)/2$, where N is the filter order. The same requirements concerning center frequency and damping factor as discussed previously for the odd-order low-pass filter apply to the high-pass case.

Appendix: Odd-Order Filters

The odd-order high-pass coefficients are summarized in Equations (A-4a), (A-4b), and (A-4c) shown below.

$$\gamma = \frac{\cos\theta_c}{1+\sin\theta_c} \qquad \alpha = (1+\gamma)/2 \qquad \text{(A-4a)}$$

$$\beta_k = \frac{1}{2}\left(\frac{1-\tfrac{1}{2}d_k \sin\theta_c}{1+\tfrac{1}{2}d_k \sin\theta_c}\right) \qquad \gamma_k = (\tfrac{1}{2}+\beta_k)\cos\theta_c \qquad \alpha_k = (\tfrac{1}{2}+\beta_k+\gamma_k)/4 \qquad \text{(A-4b)}$$

$$d_k = 2\sin\left(\frac{(2k-1)\pi}{4(M+\tfrac{1}{2})}\right) \qquad \text{(A-4c)}$$

$$\text{where,} \quad \theta_c \equiv 2\pi f_c / f_s \qquad k = 1 \ldots M$$

Coefficient formulas for the odd-order high-pass Butterworth filter

Difference Equations

In order to implement the Nth odd-order IIR filter, a set of $M+1$ *difference equations* are needed, one for the first-order section and M equations for the cascaded even-order network. Referring back to Chapters 4 and 12 for low-pass and Chapters 5 and 13 for high-pass, the odd-order filter difference equations are:

LOW-PASS:

$$y(n) = \alpha\,[x(n)+x(n-1)] + \gamma\, y(n-1)$$
$$x_1(n) = y(n)$$
$$y_1(n) = 2\{\alpha_1\,[x_1(n)+2x_1(n-1)+x_1(n-2)] + \gamma_1\, y_1(n-1) - \beta_1\, y_1(n-2)\}$$
$$x_2(n) = y_1(n)$$
$$\text{M}$$
$$x_M(n) = y_{M-1}(n)$$
$$y_M(n) = 2\{\alpha_M\,[x_M(n)+2x_M(n-1)+x_M(n-2)] + \gamma_M\, y_M(n-1) - \beta_M\, y_M(n-2)\}$$

(A-5)

DSP Filters

HIGH-PASS:
$$y(n) = \alpha\left[x(n) - x(n-1)\right] + \gamma\, y(n-1)$$
$$x_1(n) = y(n)$$
$$y_1(n) = 2\{\alpha_1\left[x_1(n) - 2x_1(n-1) + x_1(n-2)\right] + \gamma_1\, y_1(n-1) - \beta_1\, y_1(n-2)\}$$
$$x_2(n) = y_1(n)$$
$$\mathrm{M} \tag{A-6}$$
$$x_M(n) = y_{M-1}(n)$$
$$y_M(n) = 2\{\alpha_M\left[x_M(n) - 2x_M(n-1) + x_M(n-2)\right] + \gamma_M\, y_M(n-1) - \beta_M\, y_M(n-2)\}$$

where $x(n)$ is the input and $y_M(n)$ is the final filter output.

The low-pass filter coefficients in Equation (A-5) are given by Equations (A-2a), (A-2b), and (A-2c). The high-pass filter coefficients in Equation (A-6) are given by Equations (A-4a), (A-4b), and (A-4c).

Index

A

A/D converter 210, 278, 315
accumulator 43, 99
ADC 23, 37
adder 43, 52, 99
algorithm 26
aliasing 19
amplitude 19
amplitude signal 23
analog 17, 19
analog filter 64
analog response curve 64
analog signal 23
analog sinusoidal wave 19
analog-to-digital (A/D) converter 23, 177

B

band-edge 10
band-edge frequencies 82, 89, 97
band-stop filter 12, 93, 94, 100, 103, 105, 163, 164, 167, 170, 176, 207, 210, 212, 217
band-stop network 99
bandpass filter 10, 81, 90, 97, 103, 108, 113, 153, 154, 158, 160, 176, 210, 223, 226, 229, 245, 246, 249, 254, 263
bandpass network 84
bell filter 12
beta coefficient value 215
bilinear transformation 84
bilinear warping function 91
biquad section 35
boost 15, 263, 272
Butterworth filter 127, 130, 136, 140, 144, 146, 150, 155, 156, 165, 166, 171, 300, 303, 311, 336

C

cascaded band-stop 165
cascaded band-stop filter 208, 212
cascaded bandpass 165
cascaded filter 143, 148, 249
cascaded network 208
cascaded peaking filter 254
center frequency 10, 13, 81, 95, 121, 154, 167, 177, 210, 218, 221, 227, 229, 231, 247, 249, 250, 251, 269, 281
characteristic frequency 15
coefficient calculation 317
coefficient calculation block 185, 186, 189, 195, 228, 233
complex impedance 50
cones 297
crossover 297, 307, 311
crossover frequency 302, 315, 316, 317, 326, 327
crossover network 298
cut 263, 272
cutoff frequency 10, 15, 64, 71, 126, 147, 149, 180, 188, 198, 269, 275

D

D/A converter 211, 220, 243, 249, 294
DAC 36, 37
damping factor 126
decade 9
delay element 31, 43, 52, 99
development boards 1
DFT 25
difference equation 46, 56, 67, 68, 76, 105, 113, 123, 136, 150, 184, 207, 227, 271, 310, 316, 339
digital 17
digital crossover 304
digital crossover network 299
digital filter 17, 65, 75
digital filter network 43
digital frequency response 54
digital signals 1
digital-to-analog converter 36, 178
DirectX application 175, 176, 203, 214, 235, 256, 284
DirectX control interface 193, 322
DirectX implementation 175, 233, 240, 252
discrete transform 25
dithering 23
driver 297, 316

E

edge frequencies 158, 168
electromagnetic interference 205
EMI 205
EQ network 263
equalizer 179
even-order filter 335

F

farads 40, 50, 60, 79, 94
filter bandwidth 10
filter coefficients 105, 315, 326
filter cutoff frequency 43
filter damping factor 64
filter gain 210
filter order 10
filter response plot 74
filter states 105
filter transfer function 41, 70, 80, 94, 118
Finite Impulse Response 32
finite impulse response 108
finite impulse response (FIR) filter 116
FIR 32, 108, 112, 116
FIR filter 33, 34, 120
first-order digital low-pass filter 42
first-order filter 9
first-order high-pass filter 51
first-order low-pass filter 41
Fourier transform 24, 26
frequency bandwidth 82, 103
frequency response 44

G

gain 112, 227, 229
gain factor 231, 250, 275
gain response curve 64, 72, 82, 103
graphic equalizer 223, 239, 245, 248, 250, 267, 270
graphical user interface 177
GUI 177

H

harmonics 205
henrys 60, 79, 94
Hertz 40, 50, 60, 79, 94, 259
high-pass filter 10, 15, 49, 50, 139, 145, 147, 298, 312, 316, 321, 325, 328
high-pass shelving filter 183, 190, 201
hinge point 180
horns 297

I

IIR 32, 42, 76, 87, 108, 170
IIR band-stop filter 98
IIR bandpass filter 89, 160
IIR filter 32, 35, 46, 52, 150, 171
IIR peaking filter 229, 243, 271
impedance 39, 40, 50, 80
inductor leg 69
Infinite Impulse Response 32
infinite impulse response 108
input sample 201
instruction set 43, 53

L

Laplace transform 24, 26, 43, 53, 84, 99
Laplace variable 40, 50, 60, 94
least significant bit 21
limit cycling 42
Linkwitz-Riley filter 306, 308
low-pass filter 7, 9, 14, 21, 36, 39, 46, 110, 115, 116, 125, 131, 134, 139, 158, 298, 311, 316, 321, 325, 328
low-pass network 64
low-pass shelving filter 182, 190, 201
LSB 21

M

magnitude response 54, 315
main filter loop 293, 328
main initialization routine 193
maximum cutoff frequency 188, 196, 198
maximum gain boundaries 219
minimum cutoff frequency 188, 198
minimum gain boundaries 219
mks 94
mks system 79
modulo 130
multiplier 43, 52, 99

N

negative gain 263
nonrecursive filter 32
normalized center frequency 85, 100, 108, 168, 207, 251
normalized cutoff frequency 116, 187
normalized edge frequencies 168
notch filter 206
notch frequency 207
Nyquist 19, 45

Index

Nyquist frequency 45, 65, 74, 86, 87, 110, 111, 147, 168, 188, 252
Nyquist theorem 196, 232, 251, 281

O

octave 9
odd-order filter 305, 335, 337
ohms 40, 50, 60, 79, 94
out-of-band frequencies 13
output sample 201

P

parallel resistor 69
parametric equalizer 225, 268, 270
passband 8, 10
Passband Ripple 8
peaking filter 12, 14, 107, 109, 112, 176, 223, 226, 229, 244, 245, 249, 268, 278
peaking filter network 112
phase angle 44, 142, 147, 158
phase response 54, 61, 112, 143, 315
phase response curve 85
positive gain 263

Q

Q 10, 82, 108, 160
Q factor 224, 229, 231, 249, 250, 262, 263, 270, 279, 280
Q value 14
quality factor 247
quantization 21
quantization error 23

R

RCL circuit 125
RCL network 72, 93, 126
recursive filter 32, 42
requency magnitude response 61
resistive impedance 50

S

s-transfer function 84
s-domain 84
s-domain transfer function 73
s-transform 43, 53, 84
sample rate 210, 221, 227, 229, 231, 239, 247, 249, 250, 251, 275, 281, 317
sampling frequency 21
sampling rate 300
sampling theorem 19, 21, 23
second-order filter 9
shelving filter 14, 15, 115, 120, 121, 123, 176, 179, 180, 268, 269, 278, 293
signal-to-noise ratio 21
signals, digital 1
sine wave 19, 205
sinusoidal wave 19
SNR 21
Speaker crossover 299
Stop-band Attenuation 8
Stop-band Corner Frequency 8
stopband 8
summing junction 108, 116, 120, 136, 150

T

transfer function 50, 64, 94, 126, 127, 131, 140, 153, 163
transforms 24
transient distortion 298
transition band 8
Transition Band Corner Frequency 8
tweeter 298, 316, 330

U

unit circle 30
user control interface 177
utoff 10

V

vertical axis 308
voltage divider formula 70

W

waveform 19
woofer 298, 316

Z

z-domain 27, 84
z-plane 26
z-transform 26, 43, 46, 53, 64, 67, 84, 91, 99
zero axis 248

343

References

Antoniou, Andreas. Digital Filters: Analysis and Design, New York, NY: McGraw-Hill, 1974.

Berlin, Howard M. Design of Active Filters with Experiments. Indianapolis, IN: Howard W. Sams & Co., Inc., 1977.

Bogner and Constantinides, eds. Introduction to Digital Filtering. New York, NY: John Wiley & Sons, 1975.

Bohn, Dennis A., *Accelerated Slope Tone Control Equalizers*, Journal of the Audio Engineering Society, vol. 40, no. 12, December 1992.

Bohn, Dennis A., *Constant-Q Graphic Equalizers*, Journal of the Audio Engineering Society, vol. 34, no. 9, September 1986.

Brophy, J. J. Basic Electronics for Scientists. New York, NY: McGraw-Hill, 1966.

Chrysafis, A. Fractional and Integer Arithmetic Using the DSP56000 Family of General-Purpose Digital Signal Processors (APR3/D). Motorola, Inc., 1988.

Davis, Gary and Jones, Ralph, Sound Reinforcement Handbook, Second Edition, Hal Leonard Publishing Corp. 1989.

Digital Stereo 10-Band Graphic Equalizer Using the DSP56001 (APR2/D). Motorola, Inc., 1988.

Hillman, G. D. an J.E. Lane, *Real-time Determination of IIR Coefficients for Cascaded Butterworth Filters*, IEEE International Conference On Acoustics Speech and Signaling Processing, 1989.

Jackson, Leland B. Digital Filters and Signal Processing. Boston, MA: Kluwer Academic Publishers, 1986.

Lancaster, D. Active Filter Cookbook. Indianapolis, IN: Howard W. Sams & Co., Inc., 1975.

Lane, J, et. al., *Modeling Analog Synthesis with DSPs*, Computer Music Journal, Spring 1997.

Lane, J. E., *Pitch Detection using a Tunable IIR Filter*, Computer Music Journal, Vol. 14, No. 3, Fall 1990, pp. 46-59.

Lane, J., *Build 10-Band Graphic Equalizer for Stereo with DSP Technology*, Electronic Design, Vol. 36, No. 12, May 26, 1988, pp. 95-100.

Lane, J. and Garth Hillman, *Implementing IIR/FIR with Motorola's DSP56000/DSP56001*, Application Report/Motorola DSP Applications, APR7/D, 1990.

Lyons, Richard C., Understanding Digital Signal Processing, Addison Wesley Longman, Inc., 1997.

Massie, Dana C., *An Engineering study of the Four-Multiply Normalized Ladder Filter*, Journal of the Audio Engineering Society, vol. 41, no. 7/8, July/August 1993.

McClellan, J. H. and T. W. Parks. "A Unified Approach to the Design of Optimum FIR Linear-Phase Digital Filters." IEEE Trans. Circuits Systems CT-20, 1973, pp. 697-701.

Moschytzm, G. S. and P. Horn. Active Filter Design Handbook. New York, NY: John Wiley & Sons, 1981.

Oppenheim, A. V. and R. W. Schafer. Digital Signal Processing. Englewood Cliffs, NJ: Prentice-Hall, 1975.

Orfanidis, Sophocles J., *Digital Parametric Equalizer Design with Prescribed Nyquist-Frequency Gain*, Journal of the Audio Engineering Society, vol. 45, no. 6, June 1997.

Pohlmann, K.C., editor, Advanced Digital Audio, Sams, IN, 1991.

Potchinkov, Alexander, *Frequency-Domain Equalization of Audio Systems Using Digital Filters Part I: Basic of Filter Designs*, Journal of the Audio Engineering Society, vol. 46, no. 11, November 1998.

Potchinkov, Alexander, *Frequency-Domain Equalization of Audio Systems Using Digital Filters Part II: Examples of Equalization*, Journal of the Audio Engineering Society, vol. 46, no. 12, November 1998.

Proakis, John G. et al. Introduction to Digital Signal Processing. New York, NY: Macmillan, 1988.

Rabiner, L. R. and B. Gold. Theory and Application of Digital Signal Processing. Englewood Cliffs, NJ: Prentice-Hall, 1975.

Shpak, Dale J., *Analytic Design of Biquadratic Filter Sections for Parametric Filters*, Journal of the Audio Engineering Society, vol. 40, no. 11, November 1992.

Strawn, J., et al. Digital Audio Signal Processing—An Anthology. William Kaufmann, 1985.

Williams, Arthur B. Electronic Filter Design Handbook. New York, NY: McGraw-Hill, 1981.

Wilkinson, Scott, *EQ Explained*, Electronic Musician, vol.11. no. 4., April 1995.

Zolzer, Udo, Digital Audio Signal Processing, John Wiley & Sons, 1998.